How To Build Augury by Mark Alexander McClish
©2024 Mark Alexander McClish.
All rights reserved.

Published in the United States by
Curious Corvid Publishing, LLC

ISBN: 978-1-959860-32-7

Cover design and interior design by Mark Alexander McClish

Printed in the United States of America
Curious Corvid Publishing, LLC
P.O. Box 204
Geneva, OH 44041

www.curiouscorvidpublishing.com

First Edition

How to Build Augury

a Manifesto of Compassion by Mark Alexander McClish

V 1.00.1 August 27, 2023

With consultation by:

Shelby M.	Claire A.	Ashley A.
Sarah W.	Ryn L	Jessica W.
Rachel J.	Cameron J.	Alec L
Ravven W.	Daniel P.	Aurora S.
Cameron M.	Danny F.	Dais J.

Dedication
–

To the people I love, and the world we will build together
where we will no longer struggle to survive.

TABLE OF CONTENTS

LINKS

Sources & Image Credits

Official Website

Official Store

"The Current" Podcast

Are you tired of fighting to survive?

Do you feel like, no matter how hard you work to make the world around you a better place, your greatest struggle is convincing others to simply care?

Do you ever wish you could start over with a clean slate?

—

The world is plagued with problems. From overpopulation to lack of housing, food shortages, drought, pollution, rising water levels, disease, poverty, isolation, mental illness, xenophobia, not to mention greed and hatred, many people are having trouble remaining hopeful about the future. Not only is finding solutions to these problems difficult—but implementing them comes with the simple but significant obstacle that human beings are resistant to change. In the face of these overwhelming challenges, ask yourself this simple question:

What are you willing to do about it?

Will you settle for an easier life and try to outlive these problems? Or will you work to make the world a better place?

—

I propose a solution. I believe that the vast majority of problems that plague humanity are united by a single common issue: poor health—poor physical, mental, and/or moral health.

Ask yourself another question. What does the society you live in prioritize? Is it money? Maybe prestige, or ambition? Achievement? Fame? It's probably not health. Because of this, we constantly sacrifice our health for our higher priorities, and the long-term consequences of this are evident.

But what if we changed? What if we built something different?

Part One: Justification

Context

I am writing this in the year 2023. I was born, raised, and currently live in the American southeast, in the heart of a region commonly referred to as the "Bible Belt."

Three years ago, at the beginning of 2020, we experienced a global pandemic of the pneumonia-like COVID-19. The pandemic led to much of the world quarantining for several months and the deaths of around 1.7 million people by the end of the year, and a total of 676 million cases and 6,881,955 deaths as of March 10th, 2023, when Johns Hopkins University stopped collecting data[1]. America's response to the pandemic was more lax compared with other countries, and thousands of conspiracy theorists resisted containment measures, including wearing masks, social distancing, and even vaccination—a similar phenomenon to the 1918 Spanish Flu pandemic[2]. Some celebrities, including our president, used their

---

[1] Johns Hopkins University. (2023, March 10). Covid-19 map. Johns Hopkins Coronavirus Resource Center. Retrieved April 3, 2023, from https://coronavirus.jhu.edu/map.html

[2] Liang, S. T., Liang, L. T., & Rosen, J. M. (2021, May). Covid-19: A comparison to the 1918 influenza and how we can defeat it. Postgraduate medical journal. Retrieved April 3, 2023, from https://www.ncbi.nlm.nih.gov/pmc/articles/PMC8108277/

platforms to encourage their audiences to avoid vaccination, to ridicule science and spread conspiracy theories, encourage unsafe alternative treatment methods, or outright ignore the danger of the disease. Meanwhile, entire cities across the globe were shut down. Businesses and schools were closed and operated remotely, if possible. We worked from home, or were laid off. Most of us stayed in our homes, isolated, for weeks and months on end. The global stock market crashed.

A few months after quarantine started, George Floyd, a forty-six-year-old African-American man, was arrested and killed[3] after the arresting officer pinned Floyd's neck under his knee for several minutes. Tensions over racial inequality and police violence in recent years built to a movement with a simple premise: "Black Lives Matter." Opponents responded: "All Lives Matter." In the following months, protests would be held in every American state and even other countries over Floyd's death—sometimes escalating to violent riots from both sides. Some tore down Confederate statues, many of which crumpled easily because they were cheaply mass produced during the Jim Crow era. Some still fly the confederate rebel flag and call it their heritage. Ties between police and white supremacy came to light. Eventually, a new message, "defund the police," grew in popularity.

Some people painted entire streets with affirmations. Some people painted hate speech and graffiti. Our Twitter-obsessed, reality-tv show host, cult-leader president (among many others) called the disease the "China virus" and encouraged a swell of racial prejudice and hate against Asian-Americans. Nazis and white supremacists celebrated him. Handfuls of people here and there still claim the Holocaust was faked. Some of the same people very loudly believe the moon landing was faked and the Earth is flat. Some celebrities talked about racism on TV—sharing their personal experience, or if

---

[3] "Three Former Minneapolis Police Officers Convicted of Federal Civil Rights Violations for Death of George Floyd." www.justice.gov, 24 Feb. 2022, www.justice.gov/opa/pr/three-former-minneapolis-police-officers-convicted-federal-civil-rights-violations-death.

they thought the concern was blown out of proportion. Other celebrities complained about their "imprisonment" while quarantined in mansions. The value of gasoline and petroleum plummeted. More Catholic priests were convicted of child sexual abuse. Wildlife ventured into abandoned cities. My wife caught COVID while I was in another state, and we spent Christmas apart, thanking God it wasn't bad enough for her to be hospitalized.

At the beginning of 2021, we had a new president sworn in—only a few weeks following an insurrection of over 700 people who attacked the US Capitol, disrupting a joint session of Congress in the process of affirming the presidential election results[4]. Everyone called it domestic terrorism, but somehow didn't seem to be that big of a deal. Towards the middle of the year, the FDA approved the vaccine for COVID-19, the fastest a vaccine had ever been developed. Unfortunately, the preexisting "anti-vax" movement fought back, claiming a myriad of reasons from "it hasn't been around long enough to trust," to "the vaccine is actually a microchip the government is implanting in your arm to track you" (a stance they preached through their smartphones). This mindset made spreading easier for the incredibly adaptive disease, which mutated into the Delta and then Omicron variants. More people died. Even more are suffering from long-term effects like organ damage and respiratory weakness to this day. Businesses like restaurants and airlines began requiring proof of vaccination; some patrons indignantly claimed this was a violation of their personal freedom, and acquired fake vaccination documentation. The government gave us Economic Stimulus checks to stay at home (and so the economy wouldn't collapse). Middle class people were mad that lower class people got money they didn't earn. Supply chains all over the world broke down. Shelves in stores were empty. People hoarded toilet paper, and scalpers would buy it in bulk and sell it for high profit. "Essential" workers in retail and food service were forced to return to work, and were told they were heroes, though given minimal protection from

---

[4] United States Department of Justice. "One Year since the Jan. 6 Attack on the Capitol." Www.justice.gov, 30 Dec. 2021, www.justice.gov/usao-dc/one-year-jan-6-attack-capitol.

contagion and minimal wages. Some people couldn't find work or provide for themselves. The richest people in the world got significantly richer;[5] and the richest companies in the world reported record profits while paying minimum wage and raising prices of groceries. TikTok, binge-watching, and video games became increasingly popular.

2022 marked a turning point, the beginning of one era after the end of another. Russia invaded Ukraine,[6] and as I'm writing this, they still are. This sparked support for Ukraine worldwide. Tensions between the U.S. and China continued to build. Inflation got worse, but the pandemic grew more manageable, and the world opened up again.

And now, at the beginning of 2023, Russia continues its invasion of Ukraine, but for many of us, our original indignance and fear has faded. While they live and fight for their homeland in a war zone, we've gone back to normal (though our government continues to donate money we don't have[7]). We make jokes about how World War III is just around the corner. China and Russia are forming an alliance with India, as well as Brazil and South Africa (BRICS[8]), and are working on an international business currency, which could mean a serious blow to our economy[9]. America is experiencing many of the

[5] Peterson-Withorn, Chase. "How Much Money America's Billionaires Have Made during the Covid-19 Pandemic." Forbes, 30 Apr. 2021, www.forbes.com/sites/chasewithorn/2021/04/30/american-billionaires-have-gotten-12-trillion-richer-during-the-pandemic/?sh=6c2146b3f557. Accessed 3 Apr. 2023.
[6] "Ukraine - the Russian Invasion of Ukraine | Britannica." www.britannica.com/place/Ukraine/The-Russian-invasion-of-Ukraine.
[7] Fiscal Data.Treasury.Gov. "Fiscal Data Explains the National Debt." Fiscaldata.treasury.gov, 2023, fiscaldata.treasury.gov/americas-finance-guide/national-debt/. Accessed 2 June 2023.
[8] Papa, Mihaela, and The Conversation. "How Long Will the Dollar Last as the World's Default Currency? The BRICS Nations Are Gathering in South Africa This August with It on the Agenda." Fortune, 25 June 2023, fortune.com/2023/06/25/dollar-reserve-currency-brics-brazil-russia-india-china-south-africa/. Accessed 25 July 2023.
[9] Mott, Filip De. "The Dollar's Dominance Would Face a Threat Unlike Any Other from a BRICS Currency, Former White House Economist Says." Markets Insider, Insider, 25 Apr. 2023,

same factors which led to the fall of the Roman Empire[10]. We've had concentration camps[11] at our southern border for years, but no one cares—many people will become angry if you acknowledge them, and call you "bleeding heart" as if your compassion is an insult. Everything is made of plastic,[12] and microplastics are in everything.[13] The oceans have islands of plastic as large as countries. The richest individuals and largest corporations and governments pollute the air and water, fill the oceans with plastic nets[14], then guilt us into using paper straws and buying their impractical electric cars. Large petroleum companies learned the effect their pollution was having on the Earth decades ago, and suppressed the information[15] to sell more product. Tech companies manufacture billions of smartphones,

markets.businessinsider.com/news/currencies/de-dollarization-dominance-yuan-china-brics-alternative-currency-sanction-russia-2023-4. Accessed 26 Apr. 2023.

[10] Murphy, Cullen. "No, Really, Are We Rome?" The Atlantic, 11 Mar. 2021, www.theatlantic.com/magazine/archive/2021/04/no-really-are-we-rome/618075/.

[11] Pry, Alyssa. "Holocaust Survivor Ruth Bloch: Yes, the Border Detention Centers Are like Concentration Camps." The Daily Beast, The Daily Beast, 8 July 2019, www.thedailybeast.com/holocaust-survivor-yes-the-border-detention-centers-are-like-concentration-camps.

[12] Arcuri, Bronson, and Benjamin Naddaff-Hafrey. "How Everything Became Plastic." NPR.org, 28 May 2019, www.npr.org/sections/money/2019/05/28/726326078/how-everything-became-plastic. Accessed 27 Apr. 2023.

[13] Garnett, Jasmine. "Microplastics Are Everywhere, Including in Our Bodies. Here's What We Know — and Don't Know — about the Impacts." KQED, 4 Mar. 2023, www.kqed.org/news/11942296/microplastics-are-everywhere-including-in-our-bodies-what-we-know-and-dont-know-about-the-impacts. Accessed 27 Apr. 2023.

[14] Lebreton, Laurent, et al. "Industrialised Fishing Nations Largely Contribute to Floating Plastic Pollution in the North Pacific Subtropical Gyre." Scientific Reports, vol. 12, no. 1, 1 Sept. 2022, p. 12666, www.nature.com/articles/s41598-022-16529-0, https://doi.org/10.1038/s41598-022-16529-0. Accessed 5 Apr. 2023.

[15] Hall, Shannon. "Exxon Knew about Climate Change Almost 40 Years Ago." Scientific American, Scientific American, 26 Oct. 2015, www.scientificamerican.com/article/exxon-knew-about-climate-change-almost-40-years-ago/. Accessed 11 Apr. 2023.

all of which by design[16] become toxic, obsolete bricks within a few years in order to drive up demand and profit margins. Our technology, furniture, and clothes are produced using slave labor[17]—and we know it. We enjoy everyday low prices[18] throughout the country—subsidized by slavery and poverty wages to put another zero behind the owner's salary. New and more efficient ideas for technology to make our lives better are presented, celebrated, and then quickly disappear. In 2017, Daphne Caruana Galizia and Ján Kuciak discovered a conspiracy by some of the wealthiest individuals in the world to hoard wealth and dodge taxes, and they quickly disappeared[19]. "The world is built with blood, and genocide, and exploitation...the global network of capital essentially functions to separate the worker from the means of production...and every politician...protects the interests of the pedophilic[20] corporate elite."[21] Around the world there is war, terrorism, and genocide. Here at home, we won't teach children fundamental sex education—but we will teach them strategies to avoid being raped or abducted. When women are raped, sexually assaulted, and harassed, they are ridiculed and blamed, and excuses are made for their aggressors. This objectification and rape culture and the network of human trafficking rings are encouraged by an increasingly more powerful and addictive multifaceted pornography industry, which we fool

---

[16] "What Is Planned Obsolescence?" Capital One, 21 Feb. 2023, www.capitalone.com/learn-grow/money-management/planned-obsolescence/. Accessed 22 May 2023.

[17] Anti-Slavery International. "What Is Modern Slavery?" Anti-Slavery International, 2022, www.antislavery.org/slavery-today/modern-slavery/.

[18] Calero-Holmes, Brandi. "How Walmart Hurts the Economy - Small Towns - BusinessNewsDaily." Business News Daily, 2012, www.businessnewsdaily.com/2405-real-cost-walmart.html. Accessed 26 Apr. 2023.

[19] ICIJ. "The Panama Papers: Exposing the Rogue Offshore Finance Industry." ICIJ, International Consortium of Investigative Journalists, 3 Apr. 2016, www.icij.org/investigations/panama-papers/.

[20] United States District Court Southern District of New York. "Sealed Indictment 19 Crim 490." Www.justice.gov, www.justice.gov/usao-sdny/press-release/file/1180481/. Accessed 21 May 2023.

[21] Burnham, Bo. How the World Works. Self-produced, 2021, open.spotify.com/track/3ymKC2QCBeN3hMseX2hUYm. Accessed 21 May 2023.

ourselves into thinking is "sex positive." Social media industries exploit the same strategy, buying and selling our children into addiction[22] like the tobacco industry used to do, after we paved the way with child beauty pageants and sexualized outfits for high school cheerleaders. Our lives are imprisoned in our social media, to the point that we have forgotten how to be social. A generation of children[23] distorted by performative vanity are compelled to document their every thought and humiliate themselves or torment their fellow human beings for attention from strangers. We desperately try to promote ourselves through the algorithms—some find viral success and become addicted to it, but most of us can't compete with the large companies. Targeted advertisements saturate every aspect of our daily existence, even the skies[24]. We make jokes about FBI agents watching us through our phone cameras, and about Alexa being a wiretap. Companies put spikes in front of windows and arms on benches to keep homeless people away while hundreds of thousands of houses stand empty.[25] Countless people starve to death while farmers destroy mountains of food because there isn't sufficient demand to meet the supply.[26] America leads the rest of the developed world in percentage of incarcerated citizens[27]

---

[22] Andersson, Hilary. "Social Media Apps Are "Deliberately" Addictive to Users." BBC News, 4 July 2018, www.bbc.com/news/technology-44640959.

[23] Ehmke, Rachel. "How Using Social Media Affects Teenagers." Child Mind Institute, 13 Mar. 2023, childmind.org/article/how-using-social-media-affects-teenagers/. Accessed 26 Apr. 2023.

[24] Tangermann, Victor. "New Yorkers Furious at Drones Forming Huge "Candy Crush" Ad over Skyline." Futurism, 4 Nov. 2022, futurism.com/the-byte/new-yorkers-furious-candy-crush-drones. Accessed 26 Apr. 2023.

[25] United Way NCA. "Vacant Homes vs. Homelessness in the U.S." United Way NCA, 28 Mar. 2023, unitedwaynca.org/blog/vacant-homes-vs-homelessness-by-city/.

[26] Cook, Christopher D. "Farmers Are Destroying Mountains of Food. Here's What to Do about It | Christopher Cook." The Guardian, 7 May 2020, www.theguardian.com/commentisfree/2020/may/07/farmers-food-covid-19. Accessed 26 Apr. 2023.

[27] Statista. "Ranking: Most Prisoners per Capita by Country 2019 | Statista." Statista, Statista, Jan. 2023, www.statista.com/statistics/262962/countries-with-the-most-prisoners-per-100-000-inhabitants/. Accessed 26 Apr. 2023.

and school shootings.[28] Our blood pressure is some of the highest in the world[29]. Housing costs more than half of our income and landlords paint the windows shut. Teachers can't afford to buy supplies and our Secretary of Education owns a $40 million superyacht[30]. Our military spending is higher than the next 10 highest countries combined[31]. Disney has its own private local government[32], and more control over our federal government[33] than we do. Rich and famous people are publicly caught with cocaine and avoid jail time. Poor people go to jail for possessing small quantities of marijuana. Though we are not a theocracy, we boast to be a christian nation—yet our culture embodies none of the Christian virtues: love, joy, peace, patience, kindness, faithfulness, gentleness, and self control. Half of our politicians hypocritically invoke the Bible to promote their selfish gain. They push legislation on the fraudulent basis of misapplied scripture and violate our first amendment. Their fanbase considers them good Christian role models. The most well-known ministers and pastors live in gigantic mansions and preach from pulpits in auditoriums built for thousands. Look at our "good christian leaders:" they insist on their own way; they are

---

[28] World Population Review. "School Shootings by Country 2020." Worldpopulationreview.com, World Population Review, 2022, worldpopulationreview.com/country-rankings/school-shootings-by-country. Accessed 26 Apr. 2023.

[29] Centers for Disease Control and Prevention. "High Blood Pressure Facts." Centers for Disease Control and Prevention, 27 Sept. 2021, www.cdc.gov/bloodpressure/facts.htm.

[30] News, A. B. C. "Secretary of Education Betsy Devos' $40 Million Yacht Set Adrift on Lake Huron." ABC News, 26 July 2018, abcnews.go.com/Politics/secretary-education-betsy-devos-40-million-yacht-set/story?id=56840039. Accessed 3 Apr. 2023.

[31] Peter G. Peterson Foundation. "U.S. Defense Spending Compared to Other Countries." Pgpf.org, Peter G. Peterson Foundation, 11 May 2022, www.pgpf.org/chart-archive/0053_defense-comparison. Accessed 5 Apr. 2023.

[32] Russon, Gabrielle. "State-Controlled Disney World Government Board Is Ripe for "Political Mischief," Expert Says." Florida Politics - Campaigns & Elections. Lobbying & Government., 16 Feb. 2023, floridapolitics.com/archives/588572-state-controlled-disney-world-government-board-is-ripe-for-political-mischief-expert-says/. Accessed 5 Apr. 2023.

[33] "Walt Disney Co Lobbying Profile." OpenSecrets, 2021, www.opensecrets.org/federal-lobbying/clients/summary?cycle=2021&id=D000000128. Accessed 5 Apr. 2023.

envious, proud, boastful—they make a mockery of the Jesus they would have helped crucify. But the rest of the world sees their example, the squeaky wheels, and hate grow to Christianity more and more. From their example, or from personal experience with false prophets or cruel family members, many conflate Christianity not with love, but with hate. This phenomenon has led to a movement of "Satanism," which does not actually worship or even acknowledge the existence of a Satan, but exists as a metaphor in rebellion against the cultural influence of Christianity.[34]

And to top it all off: there are some for whom no amount of logic or data will be sufficient to convince them of these facts; their minds are too narrow to see anything bigger than their own predisposed experiences and perspective. They listen to only that which confirms their own biases. And so we give up trying to change their minds. We see a dozen international atrocities a week on social media, with increasing regularity and intensity, and we have grown numb. We are full of xenophobic hate and tribalism, yet simultaneously desensitized into apathy, trapped in a pit of learned helplessness.

Who can blame us? If we genuinely cared about every horrible thing we saw or heard, our anxieties would have killed us long ago. Instead we put one foot in front of the other—too exhausted to trouble ourselves with the long term, only awake enough to buy a combo from McDonald's on the way home from work to eat while we binge-watch a series on one of our many streaming services before staying up doomscrolling hours past when we needed to go to sleep. We are chronically sleep-deprived. We are malnourished and obese and willingly addicted to sugar[35] and caffeine.[36] We are depressed and anxious, suffering from unresolved trauma and chronic burn-out.

---

[34] Shapiro, Rebecca. "The Pacifist's Guide to Satanism." Columbia Magazine, 18 Oct. 2022, magazine.columbia.edu/article/pacifists-guide-satanism.
[35] Detrano, Joseph. "Sugar Addiction: More Serious than You Think | Center of Alcohol & Substance Use Studies." Alcoholstudies.rutgers.edu, alcoholstudies.rutgers.edu/sugar-addiction-more-serious-than-you-think/.
[36] Striley, Catherine L.W., et al. "Evaluating Dependence Criteria for Caffeine." Journal of Caffeine Research, vol. 1, no. 4, 1 Dec. 2011, pp. 219–225, www.ncbi.nlm.nih.gov/pmc/articles/PMC3621326/, https://doi.org/10.1089/jcr.2011.0029.

We are isolated. We are unhealthy. We have no time or energy to be creative. No time to build for the future or fix long-term problems. No time to help others. No time for self-actualization. No time for anything except following the rut. We're functioning in survival mode. Our condition is not new, but lately it has been worsening.

The point I want to make is this: we know what the problems are. I could fill this book with big problems our world is currently facing, and none of them would be a surprise to anyone in my generation. We even know how to fix most of them[37]. In fact, I grew up from childhood listening to almost every adult in my life complain about some aspect of the world we live in, and suggest some alternative to fix things. Everyone has their own idea of what's wrong with the world and how they would fix it. But they give constant complaints without any action. It seems that everyone harbors some dissatisfaction with the structure of their society. You would think this would drive us to do something about it—so why don't we?

There are a myriad of answers to this question. But the essential truth that they boil down to is, we don't fix the problems because at the end of the day, we don't genuinely want to.

I wish I could say it was something as simple as "the rich and powerful benefit from this system, and so it persists." That's true, and may account for the majority, but it's deeper than that. We perpetuate this system, too. A lot of average Americans think of themselves on the side of the rich and powerful—merely temporarily inconvenienced, but preluding success, just around the corner from the riches of the American dream. We enjoy the unreachable idea that "in America, anyone can make it" as we struggle from paycheck to paycheck. We see those below us as a greater threat to our expectation of success than those above us. I do not condemn the concept of a free market, in theory—but in practice, the ideology of capitalism has led us to an extremely selfish, competitive, and

---

[37] "Scientists Politely Remind World That Clean Energy Technology Ready to Go Whenever." The Onion, 21 May 2014, www.theonion.com/scientists-politely-remind-world-that-clean-energy-tech-18 19576507.

individualistic mentality. We were fooled into the belief of "survival of the fittest" as immutable science, despite millenia of anecdotal evidence that the core of any functioning society is cooperation.

And what about the rest of us? Unfortunately, human nature is to perpetuate a negative but familiar system rather than to risk a potentially superior but unfamiliar alternative. Sometimes it's out of apathy, sometimes because the system is so big we don't know how to even go about fixing it, but most often, I believe, we remain inactive because we are so very small and we barely have enough energy to live our lives and take care of ourselves and families. Even when we stretch the very last vestiges of our energy to its limit and work with all our might to make the world a better place, too often we attack symptoms instead of the deeper underlying issue, and therefore cut the heads off weeds and leave the roots intact for them to spring up again in the next season.

Ask yourself this. What does the culture you live in prioritize? Is it class or station? Material possessions? Maybe prestige, or ambition? Good grades and a good job? Wealth? Achievement? Followers and likes? Fame? As the old saying goes, "follow the money."

According to the US Department of Justice, the average yearly income for an individual is about $53,188[38], which is a little more than I make on a public teacher's salary. Various kinds of doctors, after spending over a decade in school, make hundreds of thousands of dollars per year. Instagram influencers also make five figures—or higher[39]. NFL players make millions[40]. Most NFL coaches

---

[38]Executive Office for the Trustees (EOUST). "United States Trustee Program Home Page." Www.justice.gov, www.justice.gov/ust/eo/bapcpa/20220401/bci_data/median_income_table.htm . Accessed 20 July 2023.
[39] Bradley, Sydney. "How Much Money Instagram Influencers Make." Business Insider, 8 Nov. 2021, www.businessinsider.com/how-much-money-instagram-influencers-earn-examples-2021-6. Accessed 5 Apr. 2023.
[40] Edmonds, Charlotte. "Here's How Much Money the Average NFL Player Makes in 2022." Nbcsports.com, NBC Sports, 29 July 2022,

make more than $10 million[41]. A-List Actors are typically compensated $15 million to $20 million per movie[42]. The CEO of Amazon receives a total yearly compensation worth over $78 billion, receiving on average over three hundred times more wealth in a single hour than an average Amazon employee makes in a year[43]. In 2018, he told the world that "the only way that I can see to deploy this much [money] is by converting my Amazon winnings into space travel[44]," and so he built a vanity space travel company— then got mad when we wouldn't let him call himself an astronaut.[45] In America, there are over 40 million people living below the poverty line[46] and over 700 billionaires, the richest of whom owns Tesla, SpaceX, and Twitter, and is worth over $180 billion[47]. Just a little further down that list is the owner of Facebook and Instagram at a

www.nbcsports.com/philadelphia/eagles/2022-nfl-salaries-how-average-players-pay-compares-stars. Accessed 4 Apr. 2023.

[41] Badenhausen, Kurt. "Highest-Paid Coaches: College Football Pay Soars but NFL Still Leads Pack." Sportico.com, Sportico, 5 Dec. 2022, www.sportico.com/personalities/executives/2022/highest-paid-coaches-2022-college-football-nfl-1234696735/. Accessed 4 Apr. 2023.

[42] Brad Tuttle. "How Much Everyone on a Movie Set Gets Paid — from the Key Grip to the Director - Business Insider." Business Insider, MONEY, 15 Dec. 2017, www.businessinsider.com/how-much-everyone-on-a-movie-set-gets-paid-2017-10. Accessed 4 Apr. 2023.

[43] Hoffower, Hillary. "We Did the Math to Calculate How Much Money Jeff Bezos Makes in a Year, Month, Week, Day, Hour, Minute, and Second." Business Insider, 9 Jan. 2019, www.businessinsider.com/what-amazon-ceo-jeff-bezos-makes-every-day-hour-minute-2018-10. Accessed 4 Apr. 2023.

[44] Clifford, Catherine. "Jeff Bezos Says This Is How He Plans to Spend the Bulk of His Fortune." CNBC, 30 Apr. 2018, www.cnbc.com/2018/04/30/jeff-bezos-says-this-is-how-he-plans-to-spend-the-bulk-of-his-fortune.html. Accessed 5 Apr. 2023.

[45] Crane, Leah. "Who Counts as an Astronaut? Not Jeff Bezos, Say New US Rules." New Scientist, 22 July 2021, www.newscientist.com/article/2285017-who-counts-as-an-astronaut-not-jeff-bezos-say-new-us-rules/.

[46] U.S. Census Bureau. "National Poverty in America Awareness Month: January 2023." Census.gov, Jan. 2023, www.census.gov/newsroom/stories/poverty-awareness-month.html. Accessed 4 Apr. 2023.

[47] "Billionaires 2021." Forbes, 2022, www.forbes.com/billionaires/. Accessed 4 Apr. 2023.

net worth of ~$76 billion. Near the bottom is Kim Kardashian, who, along with her family, is famous and idolized for being famous and idolized. People who contribute palpably nothing to our collective survival and wellbeing enjoy life on an economic scale which the rest of us can scarcely comprehend. Combined, the net worth of all the billionaires in the world combined is a little over $12 trillion—or, $12,000,000,000,000. The World Food Program estimates it would cost only $40 billion per year to end world hunger by 2030[48]. During the pandemic, farmers destroyed millions of tons of food because there was insufficient demand to support the supply[49]. To put this scale of money in perspective, consider this: if you earned $500K per year (which is about what a pediatric surgeon makes), saved all of it and didn't spend a single penny on your housing, food, clothes, entertainment, or even taxes, it would take you two thousand years to save $1 billion (without interest). The rich tell us they earned their wealth through hard work, yet the laborers we see enduring the hardest working conditions are paid in pennies. In a world where machines and tools enable us to produce incredible surplus beyond our needs, we suffer from artificial scarcity created by a tiny minority of the global population.

---

[48] World Food Program USA. "How Much Would It Cost to End World Hunger?" World Food Program USA, UN WFP, 10 Aug. 2022, www.wfpusa.org/articles/how-much-would-it-cost-to-end-world-hunger/. Accessed 4 Apr. 2023.
[49] Yaffe-Bellany, David, and Michael Corkery. "Dumped Milk, Smashed Eggs, Plowed Vegetables: Food Waste of the Pandemic." The New York Times, 11 Apr. 2020, www.nytimes.com/2020/04/11/business/coronavirus-destroying-food.html. Accessed 4 Apr. 2023.

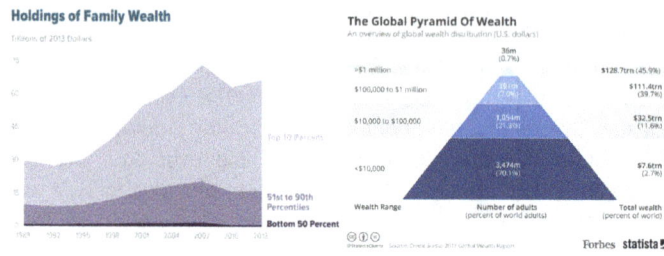

Holdings of Family Wealth
Trillions of 2013 Dollars

The Global Pyramid Of Wealth
An overview of global wealth distribution (U.S. dollars)

Congressional Budget Office, U.S. Holdings of Wealth 1989 to 2013. The top 10% of families held 76% of the wealth in 2013, while the bottom 50% of families held 1%.[50]

So, what do we prioritize? It's not health or wellbeing. Because of this, we constantly sacrifice our health (and from 2000 to 2018, 35 of the largest pharmaceutical companies reported a gross profit of $8.6 trillion) for our higher priorities, and the long-term consequences of this are evident. Americans are working longer hours than at any other time period since we started measuring, and now longer than anyone else in the industrialized world—at least 49 hours per week for over 20% of our population[51]. Heart disease— caused by high blood pressure, high cholesterol, diabetes, smoking and secondhand smoke exposure, obesity, unhealthy diet, and physical inactivity— is our leading cause of death[52]. Tobacco has been known to be deadly since the 1950s but is still legal, because our economy's reliance on the industry allowed them to lobby our government[53].

[50] "Trends in Family Wealth, 1989 to 2013 | Congressional Budget Office." Www.cbo.gov, 18 Aug. 2016, www.cbo.gov/publication/51846. Accessed 26 Apr. 2023.
[51] Schabner, Dean. "Americans: Overworked, Overstressed." ABC News, ABC News, 7 Jan. 2006, abcnews.go.com/US/story?id=93604&page=1. Accessed 5 Apr. 2023.
[52] Centers for Disease Control and Prevention. "Heart Disease and Stroke." Www.cdc.gov, National Center for Chronic Disease Prevention and Health Promotion (NCCDPHP), 7 Oct. 2020, www.cdc.gov/chronicdisease/resources/publications/factsheets/heart-disease -stroke.htm. Accessed 5 Apr. 2023.
[53] CDC. "Tobacco Industry Marketing." Centers for Disease Control and Prevention, 4 May 2018, www.cdc.gov/tobacco/data_statistics/fact_sheets/tobacco_industry/marketing/ index.htm. Accessed 5 Apr. 2023.

I could go on and on with examples, not just of poor physical health, but poor mental and spiritual health as well. We are enduring a widespread epidemic of anxiety, depression, trauma and selfishness. I posit that these issues are all effects of a deeper cause. If we want things to change, **we must address this deeper issue rather than the symptoms.**

Here is my hypothesis: I believe that the vast majority of problems that plague humanity are united by a single common issue: poor health—poor physical, mental, and/or moral health. I believe that because health is not our priority in society, we constantly sacrifice it, and commodify it, for the things we care about more. We sell ourselves, body, mind, and soul. I do not absolve the rich and powerful members of responsibility for this trend; I believe that with great power comes great responsibility, and they have created and perpetuated the demand. The people at the top take advantage of and benefit from the desperation of the people at the bottom. They have been doing this from the very beginning of the country, and even longer before that. They are possessed by the sickness of mammon. We cannot trust or rely on them to do the right thing, and we cannot wait for them to burn the world on the altar of their ego.

Ask yourself this simple question: **What are you willing to do about it**? Will you settle for an easier life and try to outlive these problems? Or will you work to make the world a better place? Not just for yourself, but for your friends, family, loved ones, children?

If you're like me, you're sick and tired of living like this. You know we could be doing better, but the people with the power to fix things won't listen—and you're pretty scared for the future. But we have the freedom and power to create the world as we see fit. Why don't we build society to serve us all, rather than benefiting the few at the expense of the many? Are we not of equal value?

Imagine a culture that prioritizes health. Imagine **your** life if health were your priority and the priority of the people around you. Would you eat differently, if different food were more readily available, or if

you had more time to cook? Would you exercise and sleep more, if you had more free time and more free spaces to inhabit without expectation of payment? But think beyond the physical—what about your mental health? What if your taxes fully paid for all the healthcare you needed, and you didn't have to worry about access to medication prescribed to you by a doctor, or your insurance denying a procedure? What if you had access to therapy, to work through trauma and pain and relieve some tension? To help you build healthy relationships? What if you had more time to spend with your loved ones, and develop those relationships? What if we were free from social media? What if we invested more in education and built our children up to be healthy and well-balanced, informed and rational from a young age? What if, when someone in our community made a bad choice, we focused on rehabilitating them instead of punishing them so badly they never fully recovered?[54] What if we focused on building a community more than turning profit? What if you didn't have to work eight hours or more per day, and had time to pursue the arts—your passions and your creativity? What if our machines did the majority of the labor, even generating a surplus, you and your people needed to survive so you weren't constantly living in a desperate struggle? What if you didn't feel the obligation to earn or justify your right to live? What if you could rest?

If those conditions changed, and your fundamental needs met, how would it change you? What would your life look like with lower stress? Would you be more relaxed, and find it easier to be patient? Would you be gentler? More merciful? Is it possible you would have more room in your heart to care about others?

---

[54] Benson, Etienne. "Rehabilitate or Punish?" Www.apa.org, American Psychological Association, July 2003, www.apa.org/monitor/julaug03/rehab. Accessed 26 Apr. 2023.

In 1943, Abraham Maslow proposed a hierarchy of human needs[55], arranged in a pyramid with survival needs at the base and self-actualization at the top. His theory posited that human motivation is not just concerned with stress reduction and survival—but also with growth and development. He suggested that once we take care of our lower, more basic needs, we are more willing and more able to pursue higher needs. I believe the same. We must care for the basic needs, not just of ourselves, but of each other, in order to be a healthier, kinder, more moral culture. We are in desperate need of a shift in fundamental perspective and priority.

Unfortunately, there is another, more legitimate obstacle to fixing the big problems. Even if we wanted to, real change takes time and great effort. The ways of a culture and the ways of a people are carved in a dirt road, in ruts that grow deeper over time, until significant deviation from the path would tear you apart. Our country was built using slave labor for hundreds of years and founded in the context of great economic, racial, and gendered division. "The way we have always done things" sets precedents, hierarchy, and bureaucracies that resist change. The problems run too deep. We

[55] Mcleod, Saul. "Maslow's Hierarchy of Needs." Simply Psychology, Simply Psychology, 21 Mar. 2023, simplypsychology.org/maslow.html. Accessed 5 Apr. 2023.

are set in our ways; too much momentum built up, running us towards our grave.

To build a society and culture that prioritizes health, we need to start over. We need a clean slate, to build **from the ground up**.

This is easier said than done, and many others before me have suggested the same. Forming a new society is not a new idea. Many groups have built small compounds or communes, sometimes successfully, and sometimes through a destructive cult. Some groups have tried to establish sovereign micronations or countries on private islands, large ships, or abandoned offshore platforms—some within another country, in their backyard. There are a few fundamental challenges with establishing a new society.

The first challenge is claiming that which has not already been claimed. The degree to which this is important is determined by how much sovereignty the society in question needs; a society at the level of a club or organization need only rent a room or building to operate out of. A society on the scale of a neighborhood or village will need multiple buildings and facilities to support a larger population living in proximity full time, and the cohabitation will overlap local ordinances. Inhabitants will be able to make and enforce some rules, but will still be subject to the larger territory's taxation, laws, and law enforcement.

To be truly sovereign (without conquering another state), a state would have to be established on "terra nullius"—land not currently controlled or occupied. In 2023, there is precious little remaining, and only the most inhospitable regions of the world that are unclaimed because they are unwanted: parts of Antarctica, the desert between Egypt and Sudan, and some regions on the Croatia-Serbia border are notable examples.

Furthermore, to support a society of people, a space of land must be able to sustain life. This means access to resources. Water is the most basic; the oldest cities in the world are situated on riverbanks or coast. Water also provides transportation, and access to food. The

space of the land itself is also important; room is needed for housing, to grow food (is there fertile soil?), for storage, for meeting places and workspaces, and so on. A society also needs some form of energy to function, for everything from powering machines to simply keeping warm. Fossil fuels are inherently inefficient[56]; does the region have an ample supply of energy in another form such as sunlight or wind, or falling water? Then there are economic resources—what materials can be produced locally, what can be sold for a profit, and what will have to be imported? Every crucial resource the land lacks must be imported, which translates to an additional burden of expense on the society. Consider transportation—is that location hard to get to? Hard to transport goods to and from? Then we must consider the effect on the local environment—how will plants and animals be displaced or affected by a new human settlement? How will the local ecosystem be affected in the long term? Our relationship is symbiotic—if we destroy nature, our own survival becomes far more difficult.

A tall order. But what if there was a place that fulfilled all of these requirements and more? A location on earth, barren and empty yet teeming with natural resources, yet unclaimed and mostly unexplored? I propose, for your consideration: the ocean.

Futurists and entrepreneurs have been looking at the possibilities of colonizing the ocean for decades, but no real attempt at an underwater facility of significant scale (beyond a few isolated examples[57]) has ever been made. The question is not whether our technology is sufficiently advanced (it is[58]), but of practicality. Underwater engineering is expensive, with too few returns on the

---

[56] Kirk, Karin. "Myths about Fossil Fuels and Renewable Energy Are Circulating Again. Don't Buy Them.» Yale Climate Connections." Yale Climate Connections, 23 Jan. 2023, yaleclimateconnections.org/2023/01/myths-about-fossil-fuels-and-renewable-energy-are-circulating-again-dont-buy-them/. Accessed 5 Apr. 2023.
[57] "Underwater Habitat." Wikipedia, 5 Feb. 2023, en.wikipedia.org/wiki/Underwater_habitat. Accessed 5 Apr. 2023.
[58] Kolitz, Daniel. "Could We Live under the Sea?" Gizmodo, 7 June 2021, gizmodo.com/could-we-live-under-the-sea-1846924986. Accessed 5 Apr. 2023.

investment to justify it. However, space travel is far, far more expensive—we pursue it not for returns on our investment, but because we really, really want to. "There's a fundamental truth to our nature—man must explore.[59]" We can colonize underwater, at far lesser expense, if we are properly motivated to.

Though inhospitable in some ways, the ocean is rich with potential. We estimate 50-80% of all life on earth inhabits the ocean. The ocean accounts for 99% of habitable space on the planet,[60] though over 90% of it remains unexplored.[61] Water, of course, is a plentiful resource—desalination is an energy-intensive but simple process which readily combines with power generation, and with a body of water comes the potential for transportation and availability of food and resources from ocean life. The motion of the water itself can be harnessed as a reliable source of energy. Sand is plentifully available, and can be used as a building component. Most of the ocean floor is empty, so the natural ecosystem need not be disrupted. In fact, using specialized techniques (which will be described in detail in Part Two), we are capable of building an underwater structure which will actively encourage ecosystems to flourish, and eventually grow into an artificial coral reef.

Furthermore, the ocean has an unconventional resource: low temperatures. Water has a high specific heat capacity, and grows cold just a few meters below the surface. Because of this, some companies like Microsoft[62] have already started pursuing underwater data centers. The highest expense for server and data center

---

[59] Astronaut David Scott, after landing on the moon in 1969.
[60] Bolles, Dana. "Living Ocean | Science Mission Directorate." Nasa.gov, 2019, science.nasa.gov/earth-science/oceanography/living-ocean. Accessed 27 Apr. 2023.
[61] Petsko, Emily. "Why Does so Much of the Ocean Remain Unexplored and Unprotected?" Oceana, 8 June 2020, oceana.org/blog/why-does-so-much-ocean-remain-unexplored-and-unprotected/. Accessed 27 Apr. 2023.
[62] Roach, John. "Microsoft Finds Underwater Datacenters Are Reliable, Practical and Use Energy Sustainably." Source, 14 Sept. 2020, news.microsoft.com/source/features/sustainability/project-natick-underwater-datacenter/.

management is cooling—but ocean water makes the process efficient and inexpensive. Digital real estate is a reliable, scalable income—and proximity to undersea communications cables means extremely fast internet connectivity. These resources can translate to consistent income on a massive scale as the demands of the digital economy grow.

This manuscript serves to present all the data and ideas I have accumulated so far, organized into a step by step plan to build a mass-capacity, self-sustaining submarine habitat and city-state, which I call Augury. Part One ends with a general overview of the organizational structure of Augury. Part Two covers the step-by-step construction plan for creating a sustainable underwater facility. Part Three details the aspects of developing a culture of health, and how the community of Augury will function, give insight into what life in Augury might look like, and address frequently asked questions.

Times are dark. The American empire is falling apart at the seams—empires historically last about 250 years, anyway, and "the Earth is littered with the ruins of empires that believed they were immortal."[63] We may be on the brink of oppressive facism or war. If it comes to war, the world as we know it may soon no longer be safe. If Albert Einstein was right[64], the surface world might soon not even be habitable. With power comes great responsibility—and with the ability to do good comes the obligation.

I am not an expert in any of the fields that will be crucial for Augury's success—I'm not an engineer or a scientist or a great leader. The ideas you'll read in this manuscript are the result of internet research over the span of a little more than a decade, but not graduate education or career experience. I'm not a war hero or a pioneer or particularly remarkable like the people who usually make history are. What I am is merely motivated to try to do better. I plan to have children one day, and my conscience demands I do everything I can,

---

[63] Camille Pagila.
[64] "I know not with what weapons World War III will be fought, but World War IV will be fought with sticks and stones." — Albert Einstein.

give every idea I have, towards building a better world for them. I'm tired of watching the people I love suffer in a prison of our own creation. We focus so much on being good descendants and honoring our past, but how often do we consider being good ancestors? In the future, I don't want to look back at this time in history, remembering how much better things used to be. I don't want our future generations to look back on their ancestors and the world they built with contempt, as we do with ours. I want to look back and remember this time as a turning point, when we started to build a better world for ourselves. We must build a more hopeful future for ourselves and our posterity. We owe it to each other. I owe it to myself and everyone I care about.

"Unless someone like you cares a whole awful lot,
Nothing is going to get better. It's not."

— *Dr. Seuss, The Lorax*

General Overview

Augury is based upon the simple idea that human beings will successfully pursue **self-optimization** in a community that effectively fulfills their needs for physical, mental, and spiritual health. The purpose of Augury is to be that community, and function as an environment that prioritizes and facilitates good physical, mental, and spiritual health.

To maintain this environment with the necessary resources and conditions, Augury must be a mass capacity, self-sustaining, sub-marine habitat. The main benefits of an underwater colony are defensibility and sovereignty, both of which are explored in detail throughout later sections. Due to the hostility of the underwater environment, primary utility and life support systems must be publicly operated as part of the local government. Most citizens of Augury will be employed in these departments associated with utility and life support in maintaining a suitable living environment for humans.

Augury's government will be divided into the same three major branches as the American government: **Judicial, Executive**, and **Legislative**. The functions of these three branches, however, are significantly different. The separation of power and system of checks and balances is a crucial similarity, however—power in Augury must be diluted and distributed to the point that no individual can consolidate an abusive degree of power. Whether an individual is originally trustworthy or not is immaterial; history has taught us over and over again that power corrupts. Augury's governing system should be organized so that officials are truly servants to the public and held accountable to the public, their responsibilities high and their compensation low, such that any office of authority would be appealing to individuals with a selfless attitude and sense of work ethic and unappealing to anyone intending to take advantage of their station.

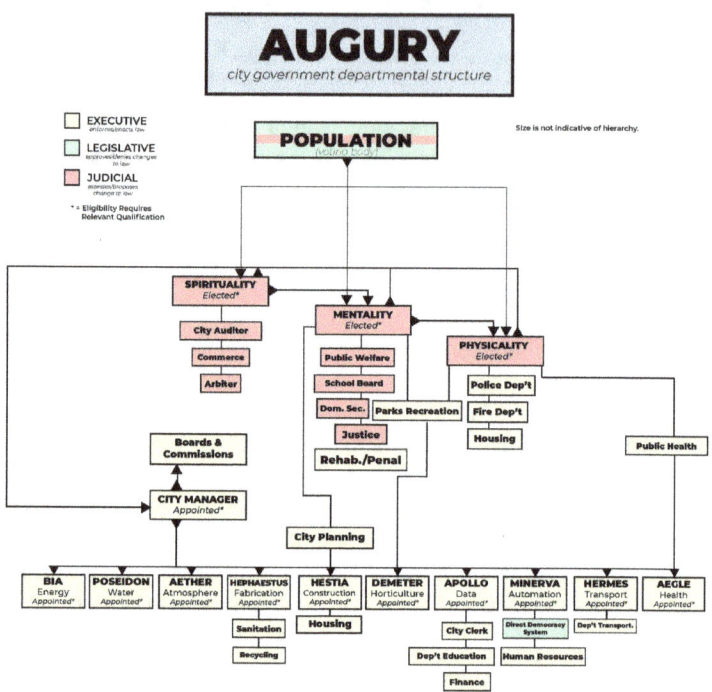

AUGURY
city government departmental structure

Executive Departments

The **Executive** branch, comprised mainly of the ten vital utility and life support departments, is responsible for performing the work necessary to keep the city running properly and keeping it habitable, anything from cleaning the air and water, expanding the city, to power generation. Note that the things people need to stay alive should never be operated for profit in Augury. The importance of these departments shouldn't be ranked as they are all crucial parts of maintaining a habitable environment (but this does not mean all government employees should receive the same compensation, as some jobs are more difficult and require more education than others).

The job of a department head is twofold: to oversee and undertake the skilled and specialized function of that department, but also to be a leader; their skills must be suited to organizing cooperation among and within teams of people, ensuring that each person's professional and personal needs are fulfilled. The unique individual needs of each person for survival is largely based on chance; the variations between these needs will be absorbed and equalized by the collective labor of the community as a whole. Teamwork and cooperation are the hallmarks of an effective society, not competition.

Furthermore, Augury should be a city which celebrates the workers, built by and for the benefit of the "blue collar" workers. That being said, governing officials should not receive preferential treatment immunity to any laws; their position must be one of public service. No abuse of their authority should ever be tolerated and they must be held to a higher moral standard than the average citizen, not less. Though the work of the executive departments covers the vast majority of the labor needed to make a society work, we must be careful to ensure that government oversight does not extend any farther over individuals personal freedom than what is necessary for the city to run smoothly. The executive departments are as follows:

BIA

Energy Generation & Storage

The Bia department is responsible for generating the energy necessary to power Augury, as well as storing and distributing that energy. Their services include maintaining power generators, wiring of buildings, transmission lines, machines, and related equipment. They may also be involved in the installation of new electrical components or the maintenance and repair of existing electrical infrastructure.

Bia works closely with other Executive departments to provide the energy they need to run their equipment and provide power to some

of the most crucial appliances: heating, water filtration, air filtration, and lights. Every department needs electricity to keep Augury running as a habitable environment. Even the core structure of the buildings themselves use power to maintain mineralization.

There's no room in a closed system like Augury for exhaust fumes, so all forms of energy generation must be clean and green, and all waste byproducts captured and reused. Nuclear fuel (like thorium salts) can be captured in a Stirling engine and used in conjunction with cold ocean water to produce reliable energy over a long period. Bia's energy production facilities should be so well-designed and automated that they could keep the city power running for decades even if they were entirely abandoned—decreasing in efficiency over time, but still functional.

An eligible candidate to head Bia would be an electrical engineer, electrician, or another individual with relevant proficiency and qualification. The Bia department will employ a team of any combination of electricians, electrical engineers, installers, technicians, and linemen as necessary.

 POSEIDON
Water Purification & Allocation

Poseidon serves the vital function of maintaining water as it flows through the city. This includes purification processes like filtering, pH balance, mineralization, and desalination. They also oversee sanitation, drainage, and the installation and maintenance of piping and fittings.

Drainage is an especially great concern in an underwater structure. Poseidon is responsible for regularly maintaining the emergency sump pumps located in every bulkhead sector, as well as the airlocks. In the event of a breach, Poseidon would contain the leak

and remedy it. Augury also has a drainage network to divert excess water down to the basement, where it can be reclaimed into plumbing or pumped outside in case of flooding.

The Poseidon department is also responsible for managing hot water supply and allocation, in which case they would work closely with Bia.

An eligible candidate to head Poseidon would be a plumber, plumbing engineer, water engineer, or another individual with relevant proficiency and qualification. The Poseidon department will employ a team of any combination of plumbers, water engineers, plumbing engineers, hydrologists, and marine biologists as necessary.

 AETHER

Atmospheric Control

In an enclosed environment like Augury, air quality is critical. The Aether department monitors the state of the interior atmosphere, maintaining balance between conditions like temperature, humidity, and pressure. They also keep a close eye on irritants, pollution, and toxic gasses, maintaining a daily Air Quality Index of never higher than 50, to ensure that even immuno-compromised individuals can breathe without irritation.

Aether's most crucial function is to balance the concentrations of oxygen and carbon dioxide by reoxygenating constantly to compensate for human respiration. Access to sufficient oxygen improves cleansing and tissue repair in the body and helps them exchange gasses more easily. Insufficient oxygen supply can result in serious health problems, but too much oxygen can cause issues as well. Reoxygenation is primarily achieved by algae and phytoplankton tanks, in which case Aether works closely with Demeter.

An eligible candidate to head Aether would be an air quality specialist, H.V.A.C. technician, or another individual with relevant proficiency and qualification. The Aether department will employ a team of any combination of H.V.A.C. technicians, air quality scientists, plumbers, electricians, and thermodynamic engineers as necessary.

 HEPHAESTUS
Manufacturing & Fabrication

There's no room for a landfill in a closed environment like Augury, so it's of great importance to minimize waste and use resources efficiently. The Hephaestus department manages manufacturing and fabrication, turning raw materials and scrap into useful things. Unlike on the surface, Augury tries to stay as far away from planned obsolescence as possible—so Hephaestus builds things to last.

They employ a variety of craftsmen to work with a great variety of materials—metals, wood, plastic, cement, ceramics, glass, acrylics, PCBs—whatever the city needs. Whenever possible, they will recycle garbage into a usable form, such as repurposing plastic litter filtered out of the ocean into building blocks. The bulk of the output of Hephaestus is construction components, which they provide for Hestia and other departments to expand, maintain, and improve the city.

Hephaestus is responsible for acquiring these raw materials, storing them, and producing supplies the city needs regularly, such as building components, textiles, and containers. But they also help the other Executive departments to do their jobs by developing or modifying specialized equipment. With cooperation and creativity, Hephaestus can find a solution to almost any problem.

Hephaestus will also maintain specialized public workshops to empower citizens of Augury to build and repair things for themselves, as well as educate industrious individuals on the tools and processes

that go into manufacturing and fabrication. These workshops include workspaces specialized for woodworking, metalworking, glass blowing, electrical work, electronic design and maintenance, textiles and sewing, and more.

An eligible candidate to head Hephaestus would be an inventor, manufacturing engineer, mechanical engineer, materials engineer, metallurgist, chemist, or another individual with relevant proficiency and qualification. The Hephaestus department will employ a team comprised of any combination of inventors, manufacturing engineers, metallurgists, chemists, materials engineers, sanitation workers, waste managers, 3D modelers, fabricators, printed circuit board designers, electronics technicians, prototypers, mechanical engineers, and technicians as necessary.

⚒ HESTIA

Construction & Housing

Augury will constantly be growing, built and maintained by Hestia. This department is not only responsible for interior structures but also the greater exterior structure of the city itself that keeps out the ocean. In the event of a breach, Hestia works closely with Poseidon to contain the water and seal the leak.

Hestia is divided into two primary categories: vital and nonvital. Vital construction includes the high pressure resistant exterior walls of the city, the load-bearing interior support struts, and the joints in between, which include entrances and airlocks. Hestia constantly maintains the structural integrity of these components to prevent breaches and failures before they happen. As new structures are grown from Biorock, Hestia tests them extensively for leaks and weak points long before anyone moves in.

Nonvital construction includes all interior structures, from housing to offices to the facilities for the other departments. This includes

41

building new structures, maintaining existing ones, and demolishing obsolete structures. Rubble is sent to Hephaestus to be recycled.

An eligible candidate to head Hestia would be a structural engineer, architect, civil engineer, or another individual with relevant proficiency and qualification. The Hestia department will employ a team of any combination of structural engineers, civil engineers, architects, technicians, surveyors, welders, metalworkers, general contractors, and laborers as necessary.

 DEMETER

Ecosystem & Horticulture

Perhaps the most widespread in functions of the Executive departments is Demeter, for the vast, categorical importance of ecosystem management. Plants provide oxygen, food, and raw materials, they are enormously beneficial for mental health, filter and moderate the atmosphere around them, are the foundation of ecosystems, and require little more than nutrients and mild radiation (sunlight and heat). And while Augury won't have room for larger animals or livestock, smaller animals like rabbits, poultry, shellfish, some insects, and small fish will be invaluable. Since Demeter is building an ecosystem from scratch, livestock will have to be upscaled in stages—first bacteria, then plants, then insects and small fish, then small animals, then larger animals.

Demeter maintains the plant life within Augury, from air-quality plants, to crops, to medicinal plants, to resource plants. They are comprised of experienced botanists, biologists, and horticulturists to maintain balance in the ecosystem to keep the plants and citizens healthy and happy.

Though trees produce significant amounts of oxygen, the oxygen production to weight ratio of plants like algae and phytoplankton is much higher; therefore they are used for the bulk of atmospheric

reoxygenation. In this instance, Demeter works very closely with Aether.

An eligible candidate to head Demeter would be an agriculturalist, biologist, botanist, horticulturist, or another individual with relevant proficiency and qualification. The Poseidon department will employ a team of any combination of gardeners, horticulturists, botanists, biologists, biochemists, chemists, agriculturists, veterinarians, and marine biologists as necessary.

 APOLLO
Networking & Data Storage

The Apollo department deals with data, whether the data is in motion or in storage. In addition to maintaining the speed and quality of the city-wide network and connection to transatlantic telecommunications cables, Apollo maintains an enormous library and data center. Apollo maintains city-wide history and records in its archives and oversees the City Clerk's office as a subdivision. Apollo's most crucial function is maintaining servers for selling digital real estate to the global market, which serves as Augury's main source of national income.

Some, but not all Minerva robots and A.I. need to store information long-term or in large quantities, such as the M.D.D.S., and do so in the Apollo data center.

An eligible candidate to head Apollo would be a data scientist, network engineer, library scientist, or another individual with relevant proficiency and qualification. The Apollo department will employ a team of any combination of network engineers, network technicians, analysts, technicians, data scientists, low voltage technicians, network security technicians, and librarians as necessary.

HERMES

 Transportation & Distribution

The Hermes department facilitates transportation, allocation, and shipping within Augury, as well as any exporting and importing. A submarine structure by nature has an inevitable degree of isolation, requiring dedicated systems in place to allow people and goods to move in and out of the city. This includes overseeing everything from submarine design and manufacture, boat traffic on the ocean surface, induction lighthouses, airlocks, and moon pools. Furthermore, the interior of Augury is designed primarily for foot traffic, but Hermes must moderate the network of elevators, public transportation, and industrial transportation. Hermes also oversees the city-wide postal system, for both local mail and international shipping.

An eligible candidate to head Hermes would be a civil engineer, traffic engineer, mechanical engineer, or another individual with relevant proficiency and qualification. The Hermes department will employ a team of any combination of traffic engineers, civil engineers, mechanical engineers, prototypers, metallurgists, welders, fabricators, programmers, and marine engineers as necessary.

MINERVA

Automation & Artificial Intelligence

The purpose of the Minerva department is to "work smarter, not harder." Minerva ensures that Augury, and especially the other Executive departments, are running as efficiently as possible. They achieve this by using technology, from simple machines to machine learning, and delegating as much work as possible to robots and artificial intelligence. Wherever a job can be done faster, better, or with less energy, Minerva is there to optimize.

One of the major responsibilities of the Minerva department is to manage the Moderated Direct Democracy System (M.D.D.S.), which makes weighing in on public policy as accessible as possible for citizens. Votes are requested, gathered, calculated, and analyzed automatically, which grants more time for humans to address greater issues. Dishonest democratic processes like gerrymandering should be rendered impossible.

Minerva works closely with every other Executive department to provide and maintain automation equipment wherever practical. This minimizes the burden of workers. Even in an advanced society, someone has to scrub the toilets—but given that robots are so easy to build, most high schools have a robotics team, the most unpleasant and crucial labor doesn't have to be performed by a human being.

An eligible candidate to head Minerva would be a programmer, computer scientist, systems architect, or another individual with relevant proficiency and qualification. The Minerva department will employ a team of any combination of programmers, computer scientists, software engineers, software developers, machine learning engineers, systems architects, data scientists and security engineers as necessary.

 AEGLE

Healthcare & Medicine

Like a repair ward for the city's population, Aegle aims to help Augury's citizens when they can't help themselves. From manufacturing life-saving substances like Insulin, replacement neurotransmitters, artificial organs, to providing emergency care and biomedical engineering, Aegle keeps the city running by addressing physical health beyond the individual's capacity to maintain it. As the primary directive of Augury is to prioritize health, Aegle works closely with the Judicial departments to monitor and encourage citizens' health.

Aegle's goal is to keep Augury's population in good health, out of compassion rather than profit. Aegle is always on the cutting edge of researching medical science to find better and more efficient ways to repair the human body.

An eligible candidate to head Aegle would be a doctor, public health director, hospital administrator, or another individual with relevant proficiency and qualification. The Aegle department will employ a team of any combination of doctors, nurses, aides, physicians, pharmacists, technicians, and therapists as necessary.

City Manager's Office

Liaison between Executive and Judicial Branches

The City Manager's office answers directly to the Judicial branch and is responsible for coordinating with all of the departments in the Executive branch. A qualified candidate has significant leadership, communication, and creative problem-solving skills, and should be familiar with a wide range of subjects to be able to unite a diverse group of workers. Human beings in the modern world usually have very specialized knowledge and skill-base, but the city manager should be a jack-of-all-trades.

The main directives of the City Manager's office is to ensure that the Executive branch departments are fulfilling the city-wide priority of health, as defined by the Judicial Heads (and to enforce their guidance), and to help the departments cooperate in any way necessary, mediating disputes, aiding communication, and assisting optimization to every extent possible. To this end, they should perform regular check-ins on the departments to assess needs.

While some officials of the city manager's office have day-to-day tasks, most upper-level officials intentionally maintain an open schedule. This is crucial to ensuring that they are ready and open to

addressing surprises, crises, and emergencies wherever they arise. In an underwater city, capable people need to be ready and able to drop what they're doing at a moment's notice and address the unexpected.

Part of the City Manager's responsibility is to maintain a connection of accessibility between Augury's people and government departments. This includes hearing citizens' concerns and addressing them, as well as keeping the people updated on what the various departments are working on or planning. Consider having the city manager appear on a regular news show to give a general update on government projects, filmed in a public place like a cafe where passersby are free to sit and watch. Transparency is important; the activities of Augury's government should be common knowledge to the citizens it answers to, just as the government's activities should be accountable to them.

Police Department

Augury's police force should be a relatively small organization of civil servants responsible for assisting and protecting the public and maintaining public order and safety. They must *rarely* use force, and only ever as a last resort. For most public disturbances, police officers work with the department of Mentality, ready to provide support if necessary to **social workers** and **therapists**. Their responsibilities mainly consist of providing help to citizens and visitors in need, connecting them with Executive and Judicial departments as needed. They will not carry a firearm, as firearms of any kind should be universally illegal in Augury. Standards and accountability for police officers, as with all positions of authority, must be exceptionally high to maintain public trust—including appropriate disciplinary action when authority is abused. To this end, all officers should be equipped with body cameras at all times.

National Guard

Though Augury should never maintain a formal military, it would be unwise to go without a defensive combat force of some kind. Consider a rapid-response "minutemen" styled militia for general emergency response, defense, and rescue, of individuals who train only one or two days per week in disciplines such as calisthenics, martial arts, weight training, and gymnastics. Though firearms and explosives of any kind must be prohibited, citizens (not exclusively members of the national guard) may choose to protect themselves with melee weapons.

Fire Department

Fire in an airtight, hyperbaric environment is catastrophic, so the Augury fire department must work quickly to extinguish any fire before it spreads. Certain areas like electrical substation rooms are outfitted with self-deploying fire extinguisher bombs as a contingency. Some smaller fires can be addressed by extinguisher turrets mounted in major rooms, concourses, and atriums, which use Minerva algorithms to automatically spray any hot spot that exceeds a certain temperature. Larger fires must be addressed in person, so firefighters will use a combination of chemical fire extinguishers and water pumped in from the ocean. If a fire grows truly out of control, the compartment can be evacuated and sealed at bulkheads, then deoxygenated. Most flammable fuels should be prohibited.

Judicial Departments

The Judicial branch, divided into the three departments of Spirituality, Mentality, and Physicality, functions as a sort of auditing system. Their responsibility is to inspect and analyze the Executive branch and the city at large to ensure that the city-wide directive of prioritizing health is optimized. They constantly search for flaws or opportunities to improve and propose solutions to keep Augury at its best. That being said, governing officials should not receive

preferential treatment immunity to any laws; their position must be one of public service. No abuse of their authority should ever be tolerated and they must be held to a higher moral standard than the average citizen, not less.

PHYSICALITY

Physical Health

The department head of Physicality is responsible for monitoring the physical health of the people of Augury. An eligible individual would have a graduate degree in medicine, nursing, health, nutrition, or a related field, and demonstrable compassion for the wellbeing of others.

The head of Physicality must lead their department in regularly monitoring conditions like air and water quality in the city, as well as tracking nutrition, sleep habits, disease, and physical activity. Physicality must ensure that the people of Augury are able to be well-fed, hydrated, sleeping well, and getting sufficient exercise. If any issues regarding the general physical health of the population arise, the department of Physicality would assess the situation, determine the cause, and address it promptly by whatever means appropriate.

Members of the Physicality department actively lead by example by being involved in their community, being involved in or organizing physical activities and other exercise, developing social programs that encourage good health, and so on.

Public Health

> "Absolutely nothing should be sold for a profit if its absence could kill you. Any modern system where people still die from lack of these resources should be dismantled."
> —IllustriousWelder87, Redditor

49

Public health in Augury, in its many forms, should be considered a collaborative effort. None of us will be able to survive at the bottom of the ocean without shelter, air, food, water, power, and healthcare. These basic needs must be provided without preferential treatment at cost, as a public service (like power and water utilities are in America); to charge more than the actual cost of the labor and materials is exploitation. Furthermore, Augury and its resources do not belong to anyone; they belong to its people equally. Poverty must never be the cause of preventable death and hardship. Workers must be compensated fairly for their labor, and there's nothing wrong with a person turning a profit from providing quality goods and services—but no one should profit from the survival of their fellow man.

As Augury is a closed system, standards for public health must be higher than they typically are on the surface. Public health is defined as "the science and art of preventing disease, prolonging life and promoting health through the organized efforts and informed choices of society, organizations, public and private, communities and individuals[65]." For Augury's Public Health Department, this means strict oversight regarding pollution, sanitization, illness, malnourishment, dehydration, and homelessness, as well as the "promotion of healthy behaviors including hand-washing and breastfeeding, delivery of vaccinations, promoting ventilation and improved air quality…suicide prevention, smoking prevention, obesity education, increasing healthcare accessibility and distribution of condoms to control the spread of sexually transmitted diseases." In short, Augury's health department must actively check to see if the people of Augury are all healthy, and help them to be if they are not.

---

[65] Wikipedia Contributors. "Public Health." Wikipedia, Wikimedia Foundation, 16 Apr. 2019, en.wikipedia.org/wiki/Public_health. Accessed 21 July 2023.

Housing

Options for housing in Augury will be more limited than on the surface. Hestia will be responsible for the construction of residential facilities of all kinds, including temporary accommodations like hotels and permanent lodgings in apartments—construction of any kind connected to the permanent structure of Augury should be performed by the department exclusively. Therefore, all residential space may have to be publicly owned (and maintained); a system to assign living space fairly will have to be designed. Things to consider include:

- Scaling up living space by family size
- Who gets relocated when newer facilities are constructed
- How residents can apply to relocate
- How residents can request specific locations (and ocean or atrium view)
- Who gets their requests fulfilled first; does anyone get preferential treatment? Are there raffles or lotteries? Is it first-come-first-serve?
- Do department heads get dedicated/special living quarters?
- How will the rental rate be calculated? Will government employees have this fee waived? Should all citizens?

Consider constructing all living units as an empty compartment, with a set size depending on family size (600 ft^2 per person, 1200ft^2 for a family of 2, 2400ft^2 for a family of 4, etc., and depending on the specific needs of the occupants). The interior (including the floorplan/layout) could then be fully customized with modular building components like wall panels, electrical and plumbing fixtures, and flooring, all of which could be easily disassembled and reused in other units when residents vacate (consider using some kind of simulator to experiment with different layouts before construction; layout could be explored in VR and export blueprints to builders). These fixtures would be purchased by the residents, so they have complete control over their living space. Consider perhaps offering a baseline "allowance" for fixtures that residents can use, and having to fund any further fixtures out of pocket. In general, occupants

should have strong rights to their space while they occupy them—no surprise inspections, no trespassing or search and seizure by government officials without a legal warrant, and protection from surprise fees or rent increase. Residents should have contractually protected rights to their living space for their entire lives unless they decide to emigrate. Residents should be grandfathered into their living space for life so that in the event of a family member's death they would not be relocated against their will.

MENTALITY

Mental, Emotional, & Social Health

The head of Mentality will be responsible for monitoring the mental, emotional, and social health of the people of Augury. An eligible individual would have a graduate degree in sociology, social work, psychology, counseling, or a related field, a demonstrable deep understanding of how the mind works, as well as patience with others.

The head of Mentality must lead their department in regularly monitoring the mental and emotional state of the people of Augury, including but not limited to stress levels, quality of life, and overall satisfaction of life. Mentality must ensure that the people of Augury are able to manage stress well, receive adequate sunlight and recreation, feel enriched and stimulated, and so on. If any issues regarding the general mental and emotional health of the population arise, the department of Physicality would assess the situation, determine the cause, and address it promptly by whatever means appropriate.

Members of the Mentality department must actively lead by example, being involved in their community. This may include being involved in or organizing community activities like meditation, support groups, and other exercises, developing social programs that encourage good mental health, and so on.

# SPIRITUALITY
Spiritual, Ethical, & Moral Health

The head of Spirituality is responsible for monitoring the spiritual health of the people of Augury. An eligible individual would have a graduate degree in ministry, philosophy, or significant experience in a related field, and a demonstrable commitment to altruism.

The head of Spirituality must lead their department in promoting kindness, selflessness, and altruism in the culture of Augury. Spirituality must ensure that the other Judicial departments are keeping the people of Augury in quality of health such that self-actualization and altruism are attainable goals. Spirituality must also address any significant trends of immorality in policy or culture. If any issues regarding the general moral health of the population arise, the department of Spirituality would assess the situation, determine the cause, and address it promptly by whatever means appropriate.

The most crucial function of the Spirituality department is to advocate for the weakest and most vulnerable members of society, to care for them, defend them from being mistreated, and help them back to self-sufficiency if possible. If there is a trend of any group of people being mistreated or downtrodden, Spirituality should determine the cause and address it by any means appropriate.

Members of the Spirituality department must actively lead by example, demonstrating good ethics and morals by serving their community, putting others before themselves, organizing charity campaigns/funds/events, and so on.

Legislative Branch

The Legislative branch, which largely relates to accepting or denying policy change, comprises the entire population of Augury. This is facilitated by the Moderated Direct Democracy System, which is monitored by Minerva. The underlying principle is that the operation of Augury, from the Judicial branch to the Executive, is ultimately under the control of its population. The Judicial and Executive branches exist to serve the best interest of the population and optimize the city in specialized fields.

Moderated Direct Democracy System

The Moderated Direct Democracy System (MDDS) is an internet-based system that makes voting as accessible as possible. The MDDS can be accessed by any citizen's personal device with an internet connection via a website or application, connecting users with their registered data with three-factor verification. This system operates in close proximity to all other internet-based municipal networking managed by Minerva and Apollo. This system removes the need for representative-based democracy and minimizes the need for politicians. Citizens may undertake law-making, policy formation, and regulation enforcement.

Any proposed change in policy (typically proposed by a head of an Executive or Judicial branch) must be put to a popular vote, and all citizens should have until a specified deadline to cast their vote. Simpler policies will have shorter deadlines than greater changes, to allow voters time to explore related data to consider consequences and alternatives.

Furthermore, the MDDS could be able to enforce conditions to balance the fairness of voting conditions including but not limited to the following examples.

The MDDS may:

- Compensate the weight of individual votes for factors such as expertise or conflict of interest.

- Not make some votes available to the entire population if the issue is only relevant to certain departments or demographics.

- Require voters to pass a comprehension quiz to demonstrate familiarity before voting on policy changes with a high estimated impact, as well as offer relevant background information concerning the issue.

- Offer a mandatory multiple-choice quiz after voting "no" on a policy change to explain their choice and request an (optional) suggested alternative for the intended goal (typically only on smaller changes)

The MDDS technicians operate as a sub-department of Minerva to design these vote modifiers and are constantly optimizing the system to be more effective and ensure any modifiers do not cause significant bias. The MDDS department itself is also open to constructive criticism by the general population, and all vote modifiers are clearly declared at the time of the vote and recorded for later reference. Certain conditions for designing votes will be universally required for all proposed policy changes, such as the following:

Proposed Policy changes:

- Must also clearly declare the intended effect

- Must declare the department that proposed them

- Must be presented for vote individually. No votes may ever be cast to apply to multiple policy changes at once.

- Must be written in language simple enough for the average citizen to understand, and phrased as simply and concisely as possible. The MDDS department will have dedicated plain-speak auditors to ensure this.

The above conditions also apply to elected officials. While most officials are appointed by the direct superior to the position, the superior must choose at least three qualified candidates for the position and then allow citizens to vote for their preferred candidate through the MDDS. For example: if the city needed to replace the head of the Bia department, the City Manager would choose at least three candidates who are willing to accept the position, have sufficient experience in electrical engineering or a related field, etc. These candidates would then 'campaign' by publicly exhibiting their qualifications in documentation not unlike a resume. Citizens would then have a certain period of time to research and vote for their preferred candidate, and current members of the Bia department would have a higher weighted vote compared to other citizens. The MDDS would then calculate the winner by popular vote and assign their role as head of Bia.

All votes, results, and contextual information would automatically be time-stamped and stored along with a current record of municipal Augury law by Apollo in a public change-tracking database repository similar to Git.

Will this system be perfect and immune to hacking or fraud? Of course not—what system is? But I don't believe it will be especially prone, and it will make democracy more accessible and accountable than ever.

Palliative Legislation

Though it will be a difficult challenge, avoiding **palliative legislation** will be one of the keys to Augury's long-term success as a balanced community. "Palliative" is a healthcare term which refers to treatment that addresses symptoms but not the underlying cause of a

condition. Making laws which address the superficial symptoms of an issue is easier, but often punishes behaviors or activities which are not inherently bad, leading to an unnecessary and unfair restriction of people's liberties. Take loitering for instance—is loitering an inherently problematic activity? Or perhaps it could be an indication of lack of public common space in the community? Or, perhaps the goal is to repel specific groups of people, perhaps those too poor to have access to or be welcome in a better public space. Alternatively, the goal may be to discourage criminal activity in a given area, in which case the criminal activity itself should be addressed, at its root. Palliative legislation addresses just the surface level of the problem and takes the shortcut, the easy way out. It punishes everyone to spite the few; instead of making fair rules, the rules are made equally unfair. It's a lazy way to address a problem temporarily and should be avoided in Augury. Communities of people are complicated, and so are the ways we address their problems, but if we give in to doing things the easy way, we let a few bad people ruin something for the majority. I will not be so lazy. It will not be easy to shift our mindset to see deeper than the surface when it comes to solving problems. It will take time. Often, the key to fixing these problems is investing in our communities on the front end rather than being reactive to behaviors we want to discourage—"an ounce of prevention is worth a pound of cure." This should not only be our approach to conflict in legislation, but in all the social issues we seek to solve.

Politicians

The concept of a politician typically evokes the idea of a representative who votes and perhaps introduces new policies on your behalf. The Moderated Direct Democracy System eliminates the need for representative politicians—the closest thing in Augury will be Department Heads and the City Manager. Like the economy, politics in Augury should be shifted away from a competitive mentality and towards cooperation. These leaders must be gracious, humble, willing to learn from and work alongside fellow candidates. Their approach should be to work alongside other officials to solve problems, rather than to compete with each other for accolades and

popularity. They should care more about the wellbeing of the city than their own personal ambition.

New Department Heads will be elected by citizens conditionally. Eligible candidates must have certain relevant credentials, experience, and education for any given role (as specified in each department description in Part 1). Consider having runner-up candidates become assistant Department Heads, to discourage bipartisanship. Political parties should be prohibited, as should donations from citizens or organizations towards political campaigns. Rather than making a spectacle of campaigning for office, positions of public office should be treated like the jobs they are. Positions in any branch of government can be held in contracts of 1 or 2 years at a time, with citizens voting to renew the contract at the end of the period or seek other candidates. Exact terms of the contract (as well as compensation and conditions for termination) should be publicly available. When candidates are being compared for a position, they should not campaign with rallies and parades—rather, they should make their qualifications publicly known, answer questions calmly and directly without manipulation, and if they take part in debates against their opponents, they should be conducted without a live audience. Candidacy for office in Augury should be a question of optimal qualification, not a performance for a popularity contest.

Part Two: Construction Phases

Note: for the sake of simplicity, the Imperative voice, Impersonal Passive voice, and Future Simple tense will be used interchangeably in Parts Two and Three.

For reference:
Sections and subsections in this manuscript can be references as follows:
"Step 1: Bia Phase 1: Site Selection"

Stages & Phases

The twenty-five building plan phases as described below can be divided into the following broader stages:

STAGE 1: SHELL

Step I. Bia Phase 1: Power Generation

Step II. Hestia Phase 1: Superstructure

Step III. Hestia Phase 2: Pressure Vessel

STAGE 2: OUTPOST

Step IV. Aether Phase 1: Local Atmosphere

Step V. Bia Phase 2: Wiring & Lighting Installation

Step VI. Poseidon Phase 1: Municipal Plumbing

Step VII. Demeter Phase 1: Composting Waste System

Step VIII. Hermes Phase 1: Basic Infrastructure

STAGE 3: FACILITY

Step IX. Aegle Phase 1: Basic Medic Facilities

Step X. Bia Phase 3: Primary Power Plant

Step XI. Hephaestus Phase 1: Basic Manufacturing Facilities

Step XII. Hestia Phase 3: Interior Architecture

Step XIII. Apollo Phase 1: Local Network

Step XIV. Apollo Phase 2: Data Center

STAGE 4: RESORT

Step XV. Aegle Phase 2: Full Medical Facilities

Step XVI. Hestia Phase 4: Finishing Work

Step XVII. Hermes Phase 2: Transportation Network

Step XVIII. Demeter Phase 2: Interior Ecosystem

Step XIX. Poseidon Phase 2: Interior Water Features

Step XX. Hephaestus Phase 2: Full Manufacturing & Fabrication

Step XXI. Minerva Phase 1: Automate Critical Systems

I. Bia Phase 1: Power Generation

Site Selection

The average depth of the ocean is over two miles deep, where the water pressure is well over 300 times the air pressure at sea level. The pressure alone makes the environment unsuitable for long-term human habitation, not only due to the extra forces exerted on structures, but for the effect of hyperbaric pressure on human physiology. Closer to the shore, in a gently sloping area called the continental shelf, the average depth is around 200 feet. An ideal site to build Augury will be anywhere from 80-120 feet deep, where the ocean pressure is only around 3-4 times the air pressure at sea level. This amount of water pressure is in the same range to what we expect to come out of our shower; offshore oil platforms are built far deeper than this. This depth is in fact still extremely shallow, compared to the average ocean depth of 2-3 miles, and humans can dive (without atmospheric suits) in deeper water with no adverse effects[66]. Still, even in relatively shallow waters, building an air-tight

---

[66] L, Peter. "How Deep Can You Dive?" Www.abyss.com.au, 19 May 2022, www.abyss.com.au/en/blog/viewpost/338/how-deep-can-you-dive. Accessed 16 Oct. 2023.

structure underwater is hard, and maintaining it is even harder. So instead—we grow it.

Using a patented technology called **Biorock** (which will be explained further in Step II), the exterior structure of Augury can be grown as a self-healing, monolithic, limestone-adjacent structure with nothing more than an electric current and a metal frame. Though the Biorock process works in any mineral-water solution, a few factors may contribute to greater success. The Atlantic ocean has several advantages—as the saltiest of the ocean basins, it will make mineral aggregation the easiest. Proximity to the American mainland, undersea communication cables, and transatlantic shipping lanes between America, Africa, and Europe will also prove beneficial. For construction and emergency purposes, it would be inadvisable to build Augury more than 20-30 miles away from a populated landmass. Access to American infrastructure and industry will likely be necessary for a few decades until Augury becomes self-sustaining—but interaction with the American economy will likely always be in demand. There are many other factors, some of which may have not even occurred to me, which will influence the suitability of a construction site for Augury. To the best of my knowledge and research, an optimal site will be 12-15 miles off the American coast. The continental shelf off the coasts of Florida up to South Carolina is especially wide, and proximity to Atlanta, Georgia (which hosts the busiest airport in the world[67]) may prove beneficial. This site boasts warm, subtropical waters, and strong currents. Alternatively, the coast of Bermuda shows promise—an established community with a tourism industry, closer to the middle of the ocean and more out of the path of hurricanes, and a smaller local government could make sovereignty easier. Various locations in the Gulf of Mexico could also work, but would be less conducive to international travel, and the Atlantic Ocean is most optimal for Biorock mineral aggregation as it is the saltiest. The site will also need to be at least 12 nautical miles

---

[67] FOX 5 Atlanta Digital Team. "Atlanta Has the Busiest Airport in the World." FOX 5 Atlanta, 6 Apr. 2023, www.fox5atlanta.com/news/hartsfield-jackson-named-busiest-airport-in-the-world. Accessed 27 Apr. 2023.

off the coast of any established nation to avoid territorial seas[68]. Unfortunately, the Exclusive Economic Zone (EEZ) of any established nation can extend 200 nautical miles from the coast, which may be inevitable if this zone entirely consumes the continental shelf. If construction on the continental shelf proves impossible or greater isolation is needed for some reason, the **Mid -Atlantic Ridge** (a huge underwater mountain range in the world, situated in the middle of the Atlantic ocean) has peaks tall enough that Augury could be constructed atop in shallow enough water to escape pressure concerns. These regions would be squarely in the high seas, and thus entirely free from any claims from an existing country. However, most of the higher peaks will be near the existing islands, which include Iceland, the Azores, St. Paul's rock, Ascension Island, St. Helena, Tristan da Cunha, Gough Island, and Bouvet Island.

Wave Power

The first and most crucial step is to arrange some flow of power. The ocean is massive, and in constant motion due to wind, tectonic activity, and the gravitational pull of the moon. As these are all renewable resources, with proper equipment, we should be able to set up a reliable power source via several methods. Engineers around the world are currently looking into harvesting wave power, and over time their endeavors grow only more efficient. There are many different forms of wave power generators, each of which specializes in a different type of wave or terrain or depth.[69]
Using these types of generators, we can generate enough power to induce mineral aggregation.

---

[68] "Maritime Zones and Boundaries." Www.noaa.gov, 20 Mar. 2023, www.noaa.gov/maritime-zones-and-boundaries. Accessed 25 July 2023.
[69] Aydingakko, Alpaslan & Mukhaini, Mohamed & Jassasi, Salim. (2016). Renewable Energy Potential in Middle East and Particularly Oman case.

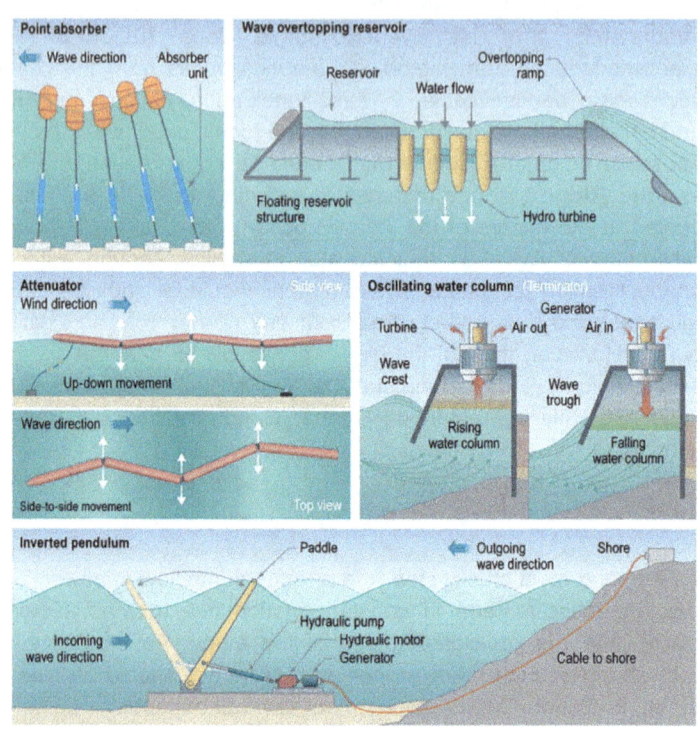

II. Hestia Phase 1: Superstructure

In 1976, an architect named Wolf Hilbertz discovered that by passing electric currents through saltwater, over time a thick layer of various minerals including limestone will accumulate on the cathode, and he patented this process under the name "Biorock[70]" (also referred to as Seacrete, Seament, and Mineral Accretion Technology). "Biorock materials are the only marine construction material that grow, get

---

[70] Patent US5543034A. https://patents.google.com/patent/US5543034A/en

stronger with age, and are self-repairing[71]." This process of mineral aggregation can form walls of limestone thick enough to withstand ocean pressure around a frame of any electrically conductive material (like rebar) using nothing but an electric current (which can be generated by the ocean itself). In addition to being inexpensive, passive, and mechanically simple, Biorock also cleans ocean water from dissolved minerals, adds hydrogen and oxygen to the water which encourages marine life, repels sharks due to its electric field, and is self-healing as long as the electric current remains active. Furthermore, Biorock is currently used to restore coral reefs because the surface of the aggregated minerals is perfect for coral growth. Therefore, the entire surface of an accreted structure will eventually grow into a coral reef, further encouraging the health of the surrounding ocean, preserving marine life by serving as an artificial marine nature preserve, facilitating tourism and marine biology (the possibilities of this will be further explored in Part 2, Step XXIII and Part 3, Economy, Tourism). Using this method, the exterior structure of Augury can be grown as a self-healing, monolithic, limestone-like structure with nothing more than an electric current and a metal frame. This framework can be manufactured on the mainland, then transported offshore, sunk, and attached to a foundational mooring platform which will be embedded deep in undersea bedrock. Using these methods we can generate a large metal-in-stone superstructure over a time period of 3-5 years (depending on electricity supplied) with relative ease.

As construction in the open ocean is dangerous, we should consider making use of drone technology to assemble the exterior framework using remotely piloted submarines. However, given our proclivity for building offshore oil platforms and the relatively low depth, this may prove unnecessarily expensive.

---

[71] "BiorockTM, Mineral Accretion TechnologyTM, SeamentTM." Global Coral Reef Alliance, 2009,
www.globalcoral.org/biorock-coral-reef-marine-habitat-restoration/.

A full team of architects and engineers will be needed to properly design the superstructure of Augury, but in the meantime, I propose the following design as a starting point:

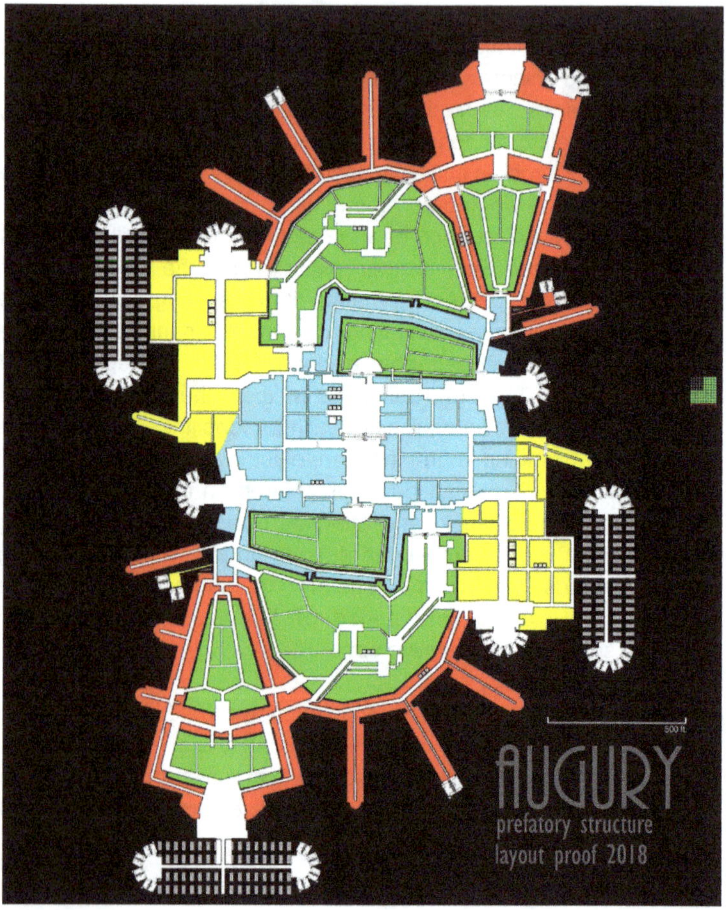

AUGURY
prefatory structure
layout proof 2018

Note that the structure is widespread and sprawling, rather than clustered under a central dome. Though gigantic domes are a common feature for underwater habitations in science fiction, they bring a risk of critical failure. To clarify, *smaller* domes are fine–in

fact, domes and arches should be used extensively in Augury's architecture, as curved shapes resist pressure best. Some atriums and arboretums can be built inside geodesic dome structures if they are strong enough. What should be avoided are designs in which **very large pockets of air with several non-watertight buildings are housed inside**, like a normal city under a dome; in this design if the dome were breached the entire interior space would be flooded. Augury should be constructed in sectors, each of which can be sealed off at bulkheads in case of a breach. To reduce pressure variation, the overall structure of Augury should be sprawling, covering horizontal space rather than rising very high. Though the stark styline on the cover art of this manuscript makes for a striking visual, it unfortunately does not represent Augury's true eventual exterior appearance. Note also that sections are color-coded (though this will be difficult to perceive if your version of this manuscript is in black and white). Office space and municipal facilities (blue) are clustered at the center, with commercial and industrial facilities (yellow) on either side (connecting to the side submarine/airlock bays). On the top and bottom are thin residential areas (red) wrapped around wide open atriums, which will primarily be green spaces. Residential areas are narrowly designed to give either views of the ocean or an atrium, to reduce claustrophobia. Main concourses are in white.

One critical feature this design lacks, however, is an absence of sharp corners— round shapes are far better at withstanding pressure than corners, so all the outermost walls should be curved and smooth. The relatively low pressure gradient and thick walls should not mandate exclusive use of spherical and cylindrical structures.

My blueprint may come with other issues I haven't noticed—I would defer to a professional firm of architects for the final design. Overall, however, Augury should be designed in the style of large hotels, airports, and shopping malls, which are made to make an enclosed structure feel open and airy (which will reduce claustrophobia). Shown on the following page is a photo (courtesy of The Arcade Providence) of a former shopping mall converted to apartments that reflects this intended style.

Consider the layout style below of an indoor shopping mall from this example of the Post Oak mall in College Station, Texas: wide walkways (made for foot traffic, but usable by small vehicles if needed) with compartmentalized spaces on the sides, easily navigable, high ceilings, usually including food courts and entertainment facilities, and ample sitting space. These buildings are designed to keep you comfortably inside as long as possible; as such, they are perfect for our purposes.

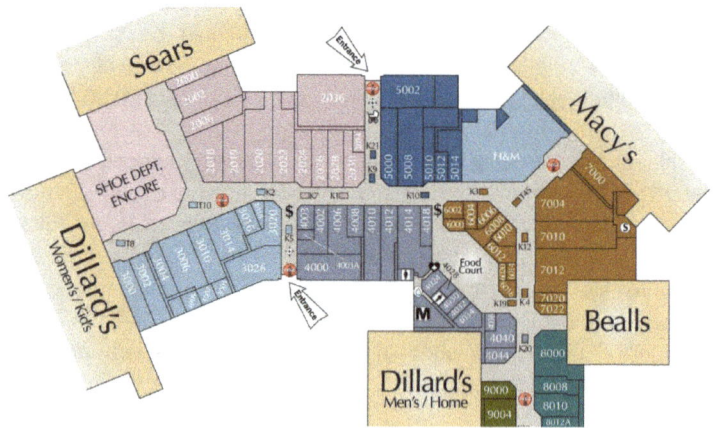

The final example below represents the ideal interior structure style for Augury: the Gaylord Opryland Resort & Convention Center in Nashville, Tennessee, which uses multiple cavernous atriums to enclose sprawling green spaces and even smaller buildings on a series of islands, bordered by hotel rooms overlooking artificial ponds and rivers. Features like these will minimize discomfort in an underwater facility. Natural light would stream in from the surface through skylights, and ample plant life will improve general mental health and air quality.

The last phase of this step should be to assemble the basic interior structure, including interior support pylons and major load-bearing structures like arches, pillars, and support framing. Building these structures as an extension of the exterior structure will make the entire superstructure stronger and establish a foundation on which to build the interior architecture in Phase 2, Step X.

Primary Induction Lighthouse

At first, the only direct access point into Augury from the surface will be the Primary Induction Lighthouse, which will be the only part of

Augury not completely submerged. Eventually, visitors, citizens, and shipping will all go through this point, and as such, it will operate as both a welcome center and a thoroughly monitored security checkpoint. At this point, it will be the center for receiving ships carrying building supplies and workers, and house the freight lifts that will carry them down into the structure. Until Augury is self-sufficient, the lighthouse will also be an umbilical structure for air, solar/wind/wave energy, satellite communications, and more. It will also house emergency egress facilities.

The lighthouse will not be directly above the main structure, but a few yards away. Though the pressure disparity will be low, the lifts that carry people down into Augury need to be slanted rather than directly vertical to allow time for decompression. Lifts carrying exclusively freight can travel faster.

Utility Shafts

Incorporating central shafts that run along the main axes of the structure could make installation and maintenance of basic utilities far more convenient. To reduce pressure variation, the overall structure of Augury should be sprawling, covering horizontal space rather than rising very high—so utility shafts could run horizontally as well as vertically. Basic utilities like electrical wiring, water pipes, ventilation shafts, data cables, and elevators/trams could be routed through these shafts.

Buoyancy Pylons

One would think building underwater is considerably more difficult than building on land—it comes with its challenges, but isn't without its advantages. One of those advantages is buoyancy. Not only does the water resistance make heavy objects easier to maneuver and fall slower, but buoyancy gives us the opportunity to build support pylons in reverse.

Usually, when you want to support an arch, beam, joist, or a ceiling within a building, you have to build columns or load-bearing walls to hold them up off the ground. If you tried to support them by suspending them from above, like with a crane, you would still have to support the crane—back to square one. If you tried to use something like a helium balloon, the buoyancy gradient with air would mean you would need 1,449 cubic feet of gas to lift every 100 pounds. Underwater, however, is a different story.

When divers need to lift a submerged object from depth, whether for equipment, rescue, salvage, or entire boats, they use something called underwater lift bags. Air is far more buoyant in water than helium is in air (less than 2 cubic feet to lift 100 pounds), and capable of lifting loads of several tons without being excessively large. Reinforced air-filled bladders could be connected from the outside to load-bearing elements of the city structure, reducing the overall dead load of the structure. This could result in an otherwise impossible architectural style, which would require far fewer supporting structures (and mitigate the challenges of larger superstructures). No building on the surface could make use of this system; it would be unique to Augury and other underwater structures. Some more precarious parts of the structure could be outfitted with emergency airbags, which would deploy in case of catastrophic collapse and prevent further damage and sinking.

These lifting bags could also be outfitted with turbines to harness power from ocean currents.[72]

---

[72] Yee, Amy. "Catching Waves and Turning Them into Electricity." The New York Times, 22 Apr. 2015, www.nytimes.com/2015/04/23/business/energy-environment/catching-waves-and-turning-them-into-electricity.html. Accessed 20 July 2023.

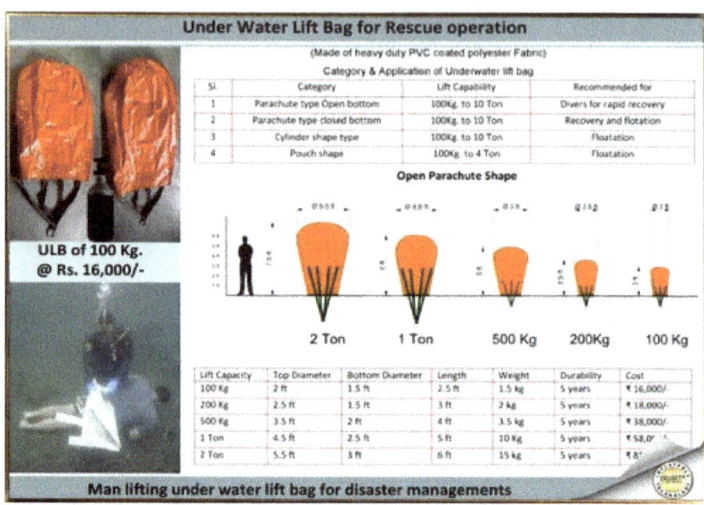

III. Hestia Phase 2: Pressure Vessel

Though gigantic domes are a common feature for underwater habitations in science fiction, they bring a risk of critical failure. To clarify, *smaller* domes are fine—in fact, domes and arches should be used extensively in Augury's architecture, as curved shapes resist pressure best. Some atriums and arboretums can be built inside geodesic dome structures if they are strong enough. What should be avoided are designs in which **very large pockets of air with several non-watertight buildings are housed inside**, like a normal city under a dome; in this design if the dome were breached the entire interior space would be flooded. Augury should be constructed in sectors, each of which can be sealed off at bulkheads in case of a breach. To reduce pressure variation, the overall structure of Augury should be sprawling, covering horizontal space rather than rising very high. Though the stark styline on the cover art of this manuscript

makes for a striking visual, it unfortunately does not represent Augury's true eventual exterior appearance.

Though Biorock is a tested, simple, and reliable technology, it is yet untested at such a scale. Doubtlessly it will be strong enough to be used as a load-bearing structure—but to my knowledge, it has never been used to construct a watertight vessel. I am confident that the structure itself can withstand the interior/exterior pressure difference (especially at such a shallow depth) but I am concerned about porosity. Will the water be able to push through the stone? If so, will the pressure inside the structure of the rock be enough to stress or fracture it? Unlikely—but we will still need to perform thorough testing at increasing scales before proceeding with inhabitation. A more likely concern is that the structure will be watertight but not airtight, and our local atmosphere may slowly bubble through and escape. However, if the interior air pressure is made equal to the exterior water pressure (which will depend on the effect of a hyperbaric environment on our health), neither will leech through the barrier. The stone will provide a structure for the water to support itself through surface tension—in fact, this phenomenon may exhibit similar qualities to the bubble formed by water anoles and act as a rebreather if oxygen and carbon dioxide can diffuse through the barrier.

If water does leach through, and if it is at a slow and controllable rate, it may be most effective to simply divert the water to the base of the structure, where it can be pumped out or somehow reused (to drive a steam turbine? Concerns about salt/mineral deposit buildup). If not, the interior surface of the exterior walls may need to be sealed with something like concrete or clay. They should not be sealed with something petroleum-based, as the harsh chemicals will undoubtedly leech into the local ecosystem.

Windows

Building a pressure vessel to withstand the aforementioned ~75 psi of water pressure is easy enough—when building with solid stone. Windows, however, add a challenge. I have not been able to calculate a minimum thickness for a window material, partially because the thickness depends on the area of the window, and partially because it would take someone with a better mind for engineering than mine to compare the variables and determine the best option. Suffice to say, ordinary glass isn't strong enough to be viable. There are a few other options each with their own advantages and disadvantages, and I'm confident that one or some combination will give us breathtaking ocean views without sacrificing much strength. From what I have found, the most promising materials (aside from the typical acrylics) are **transparent wood composites**,[73] **synthetic sapphire**,[74] and Star Trek-inspired transparent **aluminum oxynitride ceramic** (ALON).[75] Exterior windows should be layers of both rigid and flexible materials, with intent for repair in mind. Once an exterior window is damaged, we can't take it out and put in a new one without flooding the area around it. Perhaps windows can be designed in panels of a standard size, so that a special compartment tool can be laid over it to contain flooding if the panel needs replacing. Consider also including railings in front of all exterior windows, so citizens can enjoy the view while minimizing risk of any collisions.

Drainage

No matter how well we seal the structure, water will eventually, inevitably get inside the city—whether from spillage, breaches, leaks, airlocks, or just splashing. The superstructure should be designed to

---

[73] Wikimedia Contributors. "Transparent Wood Composite." Wikipedia, 28 Oct. 2021, en.wikipedia.org/wiki/Transparent_wood_composite. Accessed 22 May 2023.
[74] "Sapphire Optical Window." Www.newport.com, www.newport.com/f/sapphire-parallel-windows. Accessed 22 May 2023.
[75] Wikipedia Contributors. "Aluminium Oxynitride." Wikipedia, Wikimedia Foundation, 11 May 2019, en.wikipedia.org/wiki/Aluminium_oxynitride. Accessed 22 May 2023.

divert water along dedicated channels (perhaps coated in moss to reduce erosion) **beneath each floor** and drain into a collection cistern at the lowest point of the structure, where it can be filtered then reused or automatically removed by sump pumps. The use of **permeable pavement** as a subflooring/paving/foundational material will be a useful filter to allow water to drain to lower levels without allowing solid pollutants through.[76] We could consider utilizing separated exterior and interior shells (like a cavity wall) with a small space (of maybe 1-2 feet) in between, so that any water which leaks through the exterior wall can run down the inside surface, instead of running into the interior and causing water damage. Windows could extend through the shells. This void space would also provide greater thermal insulation from the colder ocean water, like the void space in a vacuum insulated water bottle.

Sublevel

Like the cellar/basement/crawl space of a house, the lowest level (or sublevel) of Augury's superstructure should be a utility area, used only for storage and maintenance. In this section can be stored emergency supplies and air tanks (not pure oxygen), preserved food, spare parts and building supplies, and more. The floor should be graded to direct water (which drains down from upper levels) into a system of repositories with sump pumps. The sublevel should be divided into sectors which can be sealed off in case of emergency flooding. In case of catastrophic flooding, this sublevel can be flooded entirely to protect the living areas (and for this reason the sublevel should include stations with emergency compartments or SCUBA equipment). If subterranean structures are ever needed, they can be constructed and accessed from this sublevel. If subterranean structures are ever intended to be inhabited by people (for an emergency bunker, mining, etc.), the rock should be colonized with lichen and moss with full-spectrum lighting to start

---

[76] US EPA, REG 01. "Soak up the Rain: Permeable Pavement." US EPA, 21 Aug. 2015, www.epa.gov/soakuptherain/soak-rain-permeable-pavement. Accessed 2 June 2023.

building the ecosystem. However, this is inadvisable, as isolation from any sunlight would be detrimental to the inhabitants.

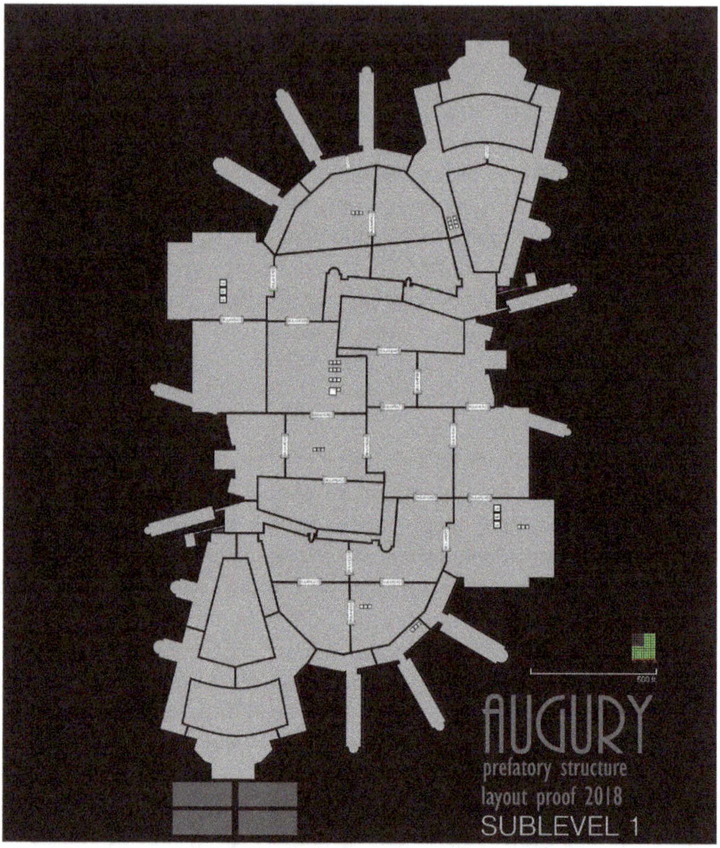

IV. Aether Phase 1: Local Atmosphere

Once the exterior structure is confirmed to be airtight, we need to start thinking about a local atmosphere in a closed system. Not only is the composition (and pollution) of the local atmosphere important, but the pressure and temperature as well, to promote good physical health. Breathing clean air can lessen the possibility of disease from stroke, heart disease, lung cancer as well as chronic and acute respiratory illnesses such as asthma. Lower levels of air pollution are better for heart and respiratory health both long- and short-term.[77]

Microalgae Reoxygenation

A single adult exhales about 2.3 pounds of carbon dioxide on an average day, an amount which would take more than 700 mature houseplants (or around 17 thousand leaves) to scrub and replace that with oxygen.[78] Though Augury will have wide-open atriums and greenspaces, 700 plants per person just isn't feasible—and carbon dioxide becomes lethal at just 5% concentration. To reoxygenate for an entire city of people, we need **cyanobacteria**, **algae,** and **phytoplankton**, which produce most of the oxygen in our atmosphere. Through natural photosynthesis, marine cyanobacteria algae can remove toxic gasses like carbon monoxide, nitrogen oxides, and sulfur oxides, with a success rate of over 98%.[79] The algae metabolizes these gasses and infuses oxygen into the filtered air, even more effectively than trees. For an even diffusion of gasses, I recommend having a piping system throughout Augury (with full spectrum LED lighting) of algae solution and strategically placed bubblers to intake air. Facilities for adding additional nutrients,

---

[77]"Breathe Easier." NIH News in Health, 28 Aug. 2018, newsinhealth.nih.gov/2018/09/breathe-easier. Accessed 20 July 2023.
[78] Candide. "How Many Plants Would It Take to Produce Enough Oxygen for One Person?" Medium, 18 Sept. 2019, medium.com/@candidegardening/how-many-plants-would-it-take-to-produce-enough-oxygen-for-one-person-7312743ed70b. Accessed 22 May 2023.
[79] Kodo, Keiun, et al. System for Purifying a Polluted Air by Using Algae. 4 July 2000, patents.google.com/patent/US6083740A/en. Accessed 22 May 2023.

removing waste, and processing purified air would be maintained by Aether. Augury's population will have to be carefully controlled so as to not overwhelm oxygenator systems, and emergency intake vents to the surface should always be maintained. Aether will also maintain a network of sensors city-wide to monitor oxygen, carbon dioxide, and carbon monoxide levels. Consider also having emergency oxygen storage tanks—though this is a fire hazard, and intake vents may be sufficient.

Hyperbaric Local Atmosphere

At the proposed ocean depth for Augury (30 to 40 meters), the water pressure is between 4 and 5 atmospheres, which is 4 to 5 times the atmospheric pressure at sea level, or 59 to 73 pounds per square inch. By maintaining a similar level of air pressure inside, Aether could reduce strain on the exterior walls and windows, and ease the transition between interior and ocean. This would be extremely beneficial to the longevity of the structure, as it would reduce leaks and breaches and therefore reduce water damage over time. However, this means that visitors to the city will have to slowly descend into the city to allow time to decompress, and the strength of the interior structure may render this unnecessary. Furthermore,

hyperbaric conditions can result in both positive[80] and negative[81] side effects over time. There are pros and cons to either choosing a hyperbaric or normal local pressure. Aegle and Physicality will need to monitor these conditions closely, as Aether will closely monitor the balance and pressure of gasses in the air.

H.V.A.C.

Heating, Ventilation, and Air Conditioning is Aether's bread and butter—and a delicate balance. At the bottom of the ocean, it's not difficult to keep the city cool. Water is a thermal insulator and has a high heat capacity, and as depth increases, temperature decreases. Heating the city is a much higher concern. Some energy can be redirected from waste heat generated by Bia, and the rest can be compensated by underground geothermal heat pumps and electric infrared heaters. Lastly, Aether must maintain proper ventilation, to ensure that heated, cleaned, and reoxygenated air is properly circulated and equally diffused among all interior spaces. Hot or cold spots are annoying—but low-oxygen spots can be hazardous and must be monitored closely. Citizens will also need to be familiar with the symptoms of oxygen deprivation, and report any instances promptly. As long as most spaces are kept open to each other, the process of diffusion will prevent this from being a concern. Individual apartments and other enclosed spaces, however, will have to be equipped with low oxygen sensors alongside smoke detectors and carbon monoxide sensors. At this stage the main ventilation network will need to be installed, along with sensors.

A **hydronic**[82] system (heating with pipes carrying hot water) embedded in floors will very likely be a more efficient method of

---

[80] Wikimedia Contributors. "Hyperbaric Medicine." Wikipedia, Wikimedia Foundation, 30 Dec. 2019, en.wikipedia.org/wiki/Hyperbaric_medicine. Accessed 22 May 2023.
[81] Wikimedia Contributors. "Oxygen Toxicity." Wikipedia, Wikimedia Foundation, 18 Nov. 2019, en.wikipedia.org/wiki/Oxygen_toxicity. Accessed 22 May 2023.
[82] a system of heating or cooling that involves transfer of heat by a circulating fluid (such as water) in a closed system of pipes.

79

heating than forced air heating. This method is more energy-efficient, especially since water has a much higher thermal mass than air, so it will retain heat and transfer it into the body of the structure, rather than just the air, more easily. This supply of water can be heated from the power plant and supplemented with evenly distributed heater tanks as necessary, and diverted into radiators where more localized heating is needed. If this system is used, Poseidon would manage city-wide heating rather than Aether. Aether would still need to maintain ventilation, however, for oxygen dispersal.

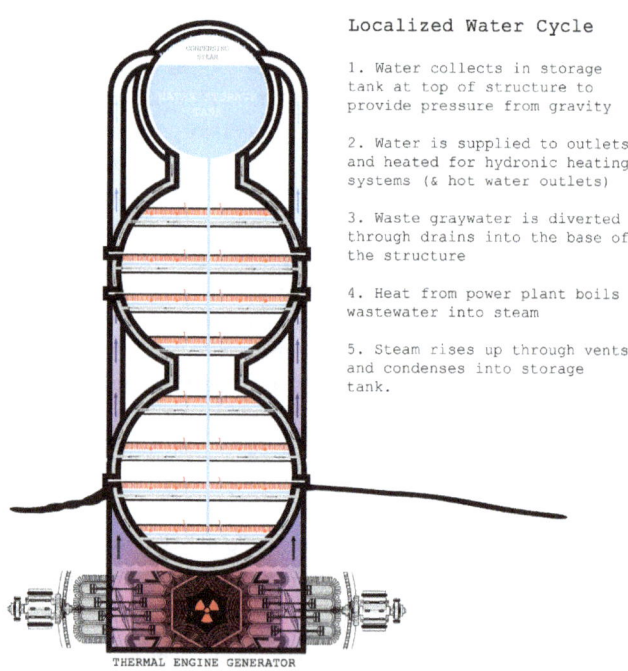

Localized Water Cycle

1. Water collects in storage tank at top of structure to provide pressure from gravity

2. Water is supplied to outlets and heated for hydronic heating systems (& hot water outlets)

3. Waste graywater is diverted through drains into the base of the structure

4. Heat from power plant boils wastewater into steam

5. Steam rises up through vents and condenses into storage tank.

THERMAL ENGINE GENERATOR

Artificial Seasons

At the bottom of the ocean, the effects of the seasons are dulled significantly. As water has a high heat capacity, water temperature at 100 feet deep will almost never grow colder than 45°F or warmer than 55°F. Rather than keeping the interior temperature at a constant temperature year-round, Aether will eventually be able to work with Minerva to shift the temperature and atmospheric conditions in larger atriums and concourses automatically to better reflect seasonal changes on the surface. This is important for ecosystem health and mental health, so citizens will have an easier time internalizing the passage of time.

Alternatively, we could make no seasons, or far milder seasons—this would require less energy overall. This could extend growing seasons and reduce seasonal depression. However, a seasonal cycle of labor (less work and more recreation during the winter) might suit our physiologies better. Consider the possibility of rotating seasons in different parts of the city, which could balance out harvests for Demeter.

Air Cleanliness

Aether could sterilize the air, killing all germs that pass through a system of air filters using UV light—but this would actually be detrimental to our health in the long run. Exposure to germs is what makes our immune systems strong, and an environment devoid of germs would make us vulnerable to anything brought down from the surface. Aether will work closely with Aegle and Physicality to keep the air clean, but not too clean. Air quality in an enclosed environment should be free of natural and artificial pollutants, at an Air Quality Index lower than 50, but never sterile. During these construction phases, thorough air filtration will be crucial to keep dust to a minimum. In medical facilities, it may be advantageous to sterilize the air as much as possible.

V. Bia Phase 2: Wiring & Lighting Installation

Due to the enclosed environment, the people of Augury won't be able to use combustible fuels of any kind. The only power source will be electrical. Before main interior structure construction begins, the entire structure should be wired thoroughly for electricity. As with all utilities, electrical wiring should be designed for longevity and ease of upkeep/maintenance (even at the expense of some aesthetic). Use of 12 gauge wire over the typical 14 should be standard for 15-20 amp circuits, to reduce risk of electrical fire. Spaces designed for industrial facilities will need to be wired with higher amp circuits, but even residential and public areas should have high-amp circuits installed periodically to run specialized machines as needed. Outlets in general should be plentiful, covered, and equipped with ground-fault circuit interrupters (GFCI) in case of water exposure.

If central utility shafts (as mentioned in Part 2: Step II) have been established, main wiring and breaker panels can be wired through them for easy maintenance.

Lighting

Many high-capacity facilities are outfitted with fluorescent or LED lighting for efficiency and cost reduction. However, these lighting types come with their side effects—strangely enough, their spectrum of blue light emissions adversely affect our mitochondria, reducing our ability to produce cellular energy.[83] Citizens may use whatever light they please in their homes, of course, but public areas and government facilities should include overhead lighting in the warmer

---

[83] Tao, Jin-Xin, et al. "Mitochondria as Potential Targets and Initiators of the Blue Light Hazard to the Retina." Oxidative Medicine and Cellular Longevity, vol. 2019, 2019, p. 6435364, www.ncbi.nlm.nih.gov/pubmed/31531186, https://doi.org/10.1155/2019/6435364. Accessed 2 June 2023.

end of the visible light spectrum. In general, using sunlight-imitative LED lighting (designed, as always, with longevity and maintenance in mind) will be better for our health. Some manufacturers are able to use full-spectrum LEDs and nanostructure filters to produce artificial light indistinguishable from sunlight.[84] If/when practical, skylights or systems of fiber optics should bring real sunlight down from the surface.

Use of lighting does have an effect on mental state, which will be described in greater detail in Part 3: Mental, Emotional, & Social Health: Sensory Overload. Cool-spectrum lighting should be used only during the day and from high overhead; otherwise, amber and gold-toned warm lighting at or near eye level should be preferred for its calming effects.[85] Exposure to blue light from cool-spectrum lighting must be minimized, as it causes a number of adverse effects including increased cell deaths and mitochondrial stress.[86]

In case of emergency power failures, lighting systems which use saltwater and ionization of magnesium (like the product WaterLight[87]) should be installed.

---

[84] Chua, Geraldine. "Construction & Architecture News." Architecture & Design, 2 Dec. 2015, www.architectureanddesign.com.au/news/led-sun-artificial-skylight-innovation-tricks-you. Accessed 2 June 2023.
[85] Colino, Stacey . "The Lighting in Your Home Could Be Affecting Your Mood." Washington Post, 11 Apr. 2023, www.washingtonpost.com/home/2023/04/11/lighting-mental-health-well-being/. Accessed 18 July 2023.
[86] Tao, Jin-Xin, et al. "Mitochondria as Potential Targets and Initiators of the Blue Light Hazard to the Retina." Oxidative Medicine and Cellular Longevity, vol. 2019, 2019, p. 6435364, www.ncbi.nlm.nih.gov/pubmed/31531186, https://doi.org/10.1155/2019/6435364. Accessed 26 Oct. 2019.
[87] Wunderman Thompson. "The Clean Energy Revolution Is Here." Wunderman Thompson, 2023, www.wundermanthompson.com/work/waterlight. Accessed 17 July 2023.

VI. Poseidon Phase 1: Municipal Plumbing

The next basic utility to be installed is plumbing. This will be installed in two stages: first, a water treatment plant to produce potable water and recycle used water. Secondly, water supply lines and fixtures in each room.

Water Treatment

We may be able to capture water from rain or surface air dehumidification, but the demands of an entire underwater facility will require us to eventually desalinate ocean water. Desalinization is a simple but energy intensive process. If a thermal engine power plant is eventually incorporated into Augury (as described in Part 2; Step X) steam may be generated then recaptured and condensed into potable water. With or without the power plant, I recommend harnessing nuclear decay heat (from a non-weaponizable material like Thorium) to desalinate water, as otherwise we would have to burn fuel (or be reliant on a surface-based sunlight capturing lens device and a system of mirrors) to generate the requisite heat. The sea salt left behind can be used as a preservative for food, exported, or even used as a building material. Wherever the salt accumulates, it should be removed regularly to reduce spread, as salt causes mechanical parts to degrade quickly. Consider protecting surfaces with sheets of flexible silicone, which can be removed and twisted to more easily remove brittle salt buildup.

Just like Aether will use algae and phytoplankton to clean the air, Poseidon will use aquatic organisms to clean the water. Some manufactured chemicals and mechanical filters will still be used where necessary, but as much as possible, water treatment should be performed by riparian plants (such as hornwort, water iris, water hyacinth, etc.)[88] as well as colonies of freshwater molluscs (especially oysters which can each filter up to 50 gallons of water per

---

[88]"12 Best Plants That Clean & Filter Pond Water 2022." Pond Informer, 30 Apr. 2022, pondinformer.com/plants-that-clean-water/. Accessed 20 July 2023.

day),[89] around artificial creeks and pools which allow the water to sit for a while, allowing pollutants to be absorbed and solids to settle, before draining down-grade into supply cisterns at the bottom of the city structure. These internal water features should follow the model of natural swimming pools,[90] in which the deepest areas are bordered by shallower shores which house the riparian filtration plants.

When the water has drained to the base of the structure, it can be funneled into a chamber with a heat source to boil. The water vapor can be vented to a cistern at the top of the structure before being condensed, so that the collected water is in a position of higher potential kinetic energy than the rest of the structure, providing water pressure for plumbing, just like a water tower.

Plumbing

To service such a large facility, water supply lines should be arranged in the "logic" layout, in which a main line connects to a multiport tee with distribution lines connected to each tee. In turn, each individual line extending from this multiport manifold provides water to all fixtures in a single or adjacent group. The system of cascading hierarchy makes plumbing systems easier to navigate for future expansion and maintenance. This layout requires far less piping material, installs faster, and leads to fewer pressure and efficiency losses. This layout does require the use of cross-linked polyethylene (or PEX) piping, which is ideal anyway— PEX (an example of a good, long-term use of a plastic) is an excellent material with many properties which make it easy to work with and ideal for plumbing,

---

[89]"Water Cleaning Capacity of Oysters Could Mean Extra Income for Chesapeake Bay Growers (Video)." NCCOS Coastal Science Website, 2 Mar. 2020, coastalscience.noaa.gov/news/water-cleaning-capacity-of-oysters-could-mean-extra-income-for-chesapeake-bay-growers-video/. Accessed 20 July 2023.

[90]Brinkley, Liv. "The Science behind Natural Swimming Pools." Grunge, 8 Apr. 2022, www.grunge.com/826223/the-science-behind-natural-swimming-pools/. Accessed 20 July 2023.

including its durability and flexibility.[91] The two big disadvantages of PEX use are a weakness to UV radiation and rodents, both of which will be less of an issue in Augury than in a typical development.

Standard hot water heaters take up a lot of space and their size and weight make them difficult to service and maintain. Tankless water heaters are preferable; though are more expensive on the front end, they are far more energy efficient as they heat water on demand without storing it. Without a storage tank, they take up very little space—a precious resource in Augury. Alternatively, larger hot water tanks or boilers could supply hot water to the whole city, and be used for heating as well. A city-wide **hydronic**[92] system routed through pipes in floors and hot-water radiators could provide more energy-efficient heating, and wouldn't dehydrate the air like forced-air heating would. The hot water will transfer heat into the thermal mass of the city structure, which means less heat will be required overall to keep the city comfortable.

Greywater

Greywater is wastewater contaminated with detergents and soap, as opposed to blackwater, which has been contaminated with sewage. Greywater will drain from sinks, showers, and washers into Augury's water systems, so available soaps and dishwasher detergents will need to be controlled to ensure no toxic substances mix with the aquatic ecosystems. Clothes washers will be automated and have ecologically-friendly detergents built-in. After filtering through layers of sand, gravel, activated charcoal, and certain plants, the filtered greywater can be reused.

Existing plants can provide eco-friendly cleaning substances. Shampoo ginger lily (*zingiber zerumbet*) is a perennial tropical plant

---

[91]PPFA Administrator. "The Benefits of PEX Plumbing Pipes - Plastic Pipe and Fittings Association." Www.ppfahome.org, 5 Aug. 2022, www.ppfahome.org/blogpost/1851733/476340/The-Benefits-of-PEX-Plumbing -Pipes. Accessed 20 July 2023.
[92] a system of heating or cooling that involves transfer of heat by a circulating fluid (such as water) in a closed system of pipes.

native to southeast Asia and Oceania which produces a clear, fragrant juice that acts as a conditioner. Crushed roots of the soapweed yucca (*yucca glauca*) produce a lather that can be used as soap or shampoo. Pineapple skins can be processed into a natural detergent[93]. Also consider atriplex roots, sapindus fruits, mojave yucca root, European soapwort root, and buffaloberry fruits. With selective breeding and genetic modification, we can further optimize plants for this purpose.

VII. Demeter Phase 1: Composting Waste System

Human urine is high in nitrogen and other nutrients, and when properly diluted, can be used as a fertilizer. Human fecal matter, on the other hand, can be dangerous if not treated carefully—but like any feces, also has potential as a fertilizer. Septic and sewer systems are space intensive and waste most nutrients—not to mention that dealing with clogs and blockages in these systems inside an enclosed, pressurized facility would be nightmarish. Instead, I propose we localize sewage treatment, and throughout Augury turn waste into resources with composting toilets.

Composting toilets do not use water (reducing our use of drinkable water) but process fecal matter using bacterial and/or fungal activity under controlled aerobic conditions. Materials like sawdust, coconut coir, or peat moss are often added after each use to improve decomposition and improve carbon-nitrogen ratio, reducing odor. Composting toilets are self contained, and do not require connection to a water source or sewage drain. These devices are growing in popularity around the world, even being used in public facilities in

---

[93] Bieber, Elizabeth McCauley, Lilian Manansala, Ryan. "Can Pineapple Skins Replace Soap?" Business Insider, 30 Mar. 2023, www.businessinsider.com/turning-pineapple-skins-into-soap-and-other-cleaning-products-2023-3. Accessed 26 Nov. 2023.

Sweden.[94] Demeter technicians could regularly (perhaps weekly) collect waste from the sealed compartment in each toilet across Augury, much like garbage collection, trading out a filled compartment with a fresh, sanitized compartment to every citizen.

The finished material can be added to Demeter's compost supply along with food waste and any other biological matter. This composting facility should use machines (compost storage containers should not be entered by humans, but should be serviced remotely by drones when necessary) to turn the decaying biological matter and monitor moisture, gas buildup, and temperature, with augers to reduce clumping. This storage facility should connect to dispensary shafts to be used by Demeter gardeners to enrich soil. Earthworms may also be included in the compost if temperature buildup is a concern, or to further process the material. Gas buildup could be pumped out from the top of the composting storage container and used as biogas fuel. If excess compost is generated beyond our ability to use, it will not harm the exterior ocean ecosystem after treating and can be dumped (far away and down-current from reefs), but this should be considered a last resort.

Toilet paper is resource and energy intensive to produce. Instead, consider using leaves of certain plants—*Coleus barbatus*, also known as *Plectranthus barbatus,* is a tropical perennial plant which produces the medicinal substance forskolin. The leaves are soft and have a minty aroma, and can be used (with or without some additional processing) in lieu of toilet paper. Aster, magnolia, thimbleberry, lamb's ear, cabbage, banana, and grape also have suitable leaves.

---

[94] Carini, Frank. "Composting Toilets Relieve Need to Flush Clean Water." EcoRI News, 19 July 2018, ecori.org/2018-7-19-composting-toilets-relieve-need-to-flush-clean-water/. Accessed 20 July 2023.

VIII. Hermes Phase 1: Basic Infrastructure

As workers can't simply drive around to different parts of the building, as soon as the interior of the structure is habitable, the next step should be to establish some system of transportation. Most likely, this will be a system of small electric vehicles to transport people, supplies, and equipment. Locations for temporary and permanent storage can be established, to facilitate an inventory and allocation system. Concourses can be also established with painted lines at this time, and emergency evacuation routes and muster stations should be established and drilled. Augury will never facilitate infrastructure for internal combustion engines; consideration must be made for foot and small electric vehicle traffic flow. Augury should be laid out on an easily-navigable grid. Connectivity is important, so passageways should be connected wherever possible so that the shortest distance between any two points is accessible. Elevators should also be installed at this stage. Airlocks and moon pools for passage out into the sea should be installed, along with a network to remotely monitor them.

Depending on the scale of the initial structure, this point might be a good opportunity to start a public transportation system. A dedicated train spanning the circumference of the city will save significant time and resources for transporting both people and materials. Cars for both freight and passengers should be included. At some point (if not now), a robust public transportation system will be crucial. There will be no room for freeways or cars in Augury, no personal vehicles much larger than bicycles, and public transportation like trains and trams are far more efficient than any other form of ground transportation.[95]

---

[95] "Rail." International Energy Agency, 11 July 2023, www.iea.org/energy-system/transport/rail. Accessed 20 July 2023.

Postal System

An important function of the Hermes department will be material/freight/cargo allocation. At this phase, this will mainly relate to delivery of building materials. Once the city is populated, it will translate into a postal system. Establishing an efficient allocation infrastructure at this phase will benefit the city in perpetuity. These systems should require as little human interaction as possible—omni-directional conveyor belts can scan barcode labels to send most package sizes all over the city automatically.

IX. Aegle Phase 1: Basic Medic Facilities

Once the city is wired with electricity, a local atmosphere is established, and workers and materials are being transported around the city to continue construction, workers will be able to live full time inside the structure while it is being built. As soon as the structure is used for full-time habitation, it's imperative that we establish a medical facility. Safety should be a top priority—many large construction projects throughout history have cost a high death toll. With rigorous safety and emergency protocols, use of remotely controlled vehicles when necessary, and careful pacing, this should not be the case with the construction of Augury. Regardless, the ocean is a harsh and unusual construction environment, and the safety and wellbeing of builders must be a top priority. Decompression sickness will be a concern, as well as the generalized hazards of an active construction site. The medical facility should have a central location that is easily reached, with dedicated infrastructure established by Hermes to minimize the time before a patient can receive treatment.

A note about healthcare culture—on the surface, medical professionals of all kinds being constantly overworked and stressed, in their education as well as in their career, is normalized. This must stop in Augury. Stress degrades the quality of their incredibly valuable work—work which should be treated with the utmost

appreciation. Nurses, pharmacists, EMTs, therapists, doctors, and anyone else who works to care for others should be given as much support as they need. Though accountability for malpractice is paramount, abuse or disrespect of these workers by patients is completely intolerable. Additionally, the more they are stressed and exhausted, the more this normalizes stress and exhaustion for them. This will shift their perception of a normal level of stress, which makes it more difficult for them to empathize with patients suffering from similar symptoms.

X. Bia Phase 3: Primary Power Plant

Though wave power is plentiful, our current technology is relatively inefficient at harvesting it—not to mention that wave power requires significant facilities for energy storage, to balance calm days and stormy days. Additionally, biofouling and the harsh environment of the ocean makes maintenance difficult. It is unlikely that wave power will be a viable power source for an entire city. Once the main facility of Augury is set up and liveable, we can pursue another, superior power source: nuclear.

I feel the need now to devote a paragraph in defense of nuclear power, to ward off any visions of mushroom clouds, fallout shelters, and radioactive mutants. Despite certain historical events, "nuclear power is a much safer energy source than fossil fuels."[96] In the past, poor regulation and negligence led to environmental disasters like those at Chernobyl and Three Mile Island. This, combined with propaganda and the nuclear threat of the Cold War, has turned public opinion against nuclear power. But I would posit, handled responsibly, nuclear power is our best source of energy on earth. I could continue in defense of nuclear fission reactors, but for Augury, I

---

[96] Boeck, Helmuth. "Why Nuclear Power Is Safer than Ever." GIS Reports, Geopolitical Intelligence Services AG, 1 Feb. 2022, www.gisreportsonline.com/r/nuclear-energy-safe/. Accessed 22 May 2023.

propose harnessing nuclear energy from an even simpler source: nuclear decay heat.

Nuclear Decay Heat

Augury should never manufacture nuclear weapons nor ever maintain a nuclear weapon stockpile. Any weapons manufacturing and stockpiling should be exclusively for the sake of domestic defense.

Here are a few facts. Radioactive material is abundant on Earth.[97] In fact, radioactive isotopes account for about half of the Earth's internal heat.[98] Nuclear waste is also abundant—"more than a quarter million metric tons of highly radioactive waste sits in storage near nuclear power plants and weapons production facilities worldwide, with over 90,000 metric tons in the US alone."[99] Furthermore, nuclear waste doesn't just come from nuclear reactors, because coal ash is also radioactive because it's full of thorium, uranium, and radon[100]. That nuclear waste sits in storage facilities, usually underground in a desert, guarded by haunting signs to warn future generations of the ongoing danger. That fuel will remain radioactive—and thermally hot— for anywhere from a few decades to thousands of years. If not actively cooled, the decay heat of radioactive waste grows "hotter and hotter as more and more heat is [generated], so the temperature will rise higher and higher…like a furnace that just never stops burning. So there is really no limit as to how hot the surroundings

---

[97] World Nuclear Association. "Uranium Supplies: Supply of Uranium." World Nuclear Association, Sept. 2021, world-nuclear.org/information-library/nuclear-fuel-cycle/uranium-resources/supply-of-uranium.aspx. Accessed 22 May 2023.
[98] Preuss, Paul. "What Keeps the Earth Cooking?" News Center, 17 July 2011, newscenter.lbl.gov/2011/07/17/kamland-geoneutrinos/. Accessed 22 May 2023.
[99] Jacoby, Mitch. "As Nuclear Waste Piles Up, Scientists Seek the Best Long-Term Storage Solutions." Chemical & Engineering News, American Chemical Society, 30 Mar. 2020, cen.acs.org/environment/pollution/nuclear-waste-pilesscientists-seek-best/98/i12. Accessed 22 May 2023.
[100] Hvistendahl, Mara. "Coal Ash Is More Radioactive than Nuclear Waste." Scientific American, Scientific American, 13 Dec. 2007, www.scientificamerican.com/article/coal-ash-is-more-radioactive-than-nuclear-waste/. Accessed 1 Sept. 2023.

can get if that heat is allowed to keep building up."[101]   If that just sounds scary, you're not seeing the possibilities. We have in our possession a material that emits limitless heat without input energy or fuel for incomparably long periods of time. This material is extremely dangerous if not contained, of course, but if it *were* properly contained, that heat could be used as a massive, scaleable, reliable power source. With layered containment of combined materials like glass, concrete, lead, and steel, and insulated with a vacuum layer and materials like aerogel, with redundant sensors, failsafes, and emergency procedures, we could have as close to a perpetual engine as humanity is capable of creating. Nuclear and mechanical engineers will be necessary to design the generator properly—but consider this endless heat source powering a simple steam turbine (which we use to power our cities on land, heated by fossil fuels[102]). Molten Thorium Salt Reactors are also becoming a safe viable alternative use for radioactive material.[103] Or, if efficiency allows, we can consider harnessing another plentiful resource of the ocean and use water's unusually high heat capacity[104] to produce a thermal gradient with the fuel and drive a Stirling engine.

Stirling Engine

Invented by Robert Stirling in 1816, a Stirling engine uses the "Stirling cycle," which harnesses motion from a temperature gradient between a heat source and a heat sink, with a regenerator in the

[101] Edwards, Ph.D., Gordon. "Do Radioactive Wastes Produce Heat?" Canadian Coalition for Nuclear Responsibility., 12 July 2011, www.ccnr.org/Radwaste_Heat.html. Accessed 22 May 2023.
[102] US Energy Information Administration. "How Electricity Is Generated - U.S. Energy Information Administration (EIA)." www.eia.gov, 9 Nov. 2022, www.eia.gov/energyexplained/electricity/how-electricity-is-generated.php. Accessed 22 May 2023.
[103] "Molten Salt Reactors - World Nuclear Association." World Nuclear Association, 2018, world-nuclear.org/information-library/current-and-future-generation/molten-salt-reactors.aspx. Accessed 22 May 2023.
[104] Water Science School. "Specific Heat Capacity and Water | U.S. Geological Survey." Www.usgs.gov, 6 June 2018, www.usgs.gov/special-topics/water-science-school/science/specific-heat-capacity-and-water. Accessed 22 May 2023.

middle to balance forces. They have the highest theoretical efficiency of any thermal engine but...a low output power to weight ratio, therefore Stirling engines of practical output tend to be large[105]. Stirling engines are slow to start (irrelevant, if we intend to have it constantly running) but have many advantages: A Stirling engine is a closed system, so the gasses used inside never leave the pistons. There are no exhaust valves and no explosions taking place, which makes them much safer than a combustion engine. The gasses inside the engine don't even need to be much higher pressure than standard, so lighter metals like aluminum can be used, and there's no risk of explosion.[106] They are mechanically simple, and therefore easy to repair and maintain. They produce very little noise and vibration, so they will not disturb citizens or ocean life. Stirling engines also boast a relatively high thermal efficiency—usually around 40%.[107] Using nuclear waste as fuel also means the generator will very infrequently need refueling.

[105] Saindon, Luke. "Thermodynamic Theory of the Ideal Stirling Engine." Blog.mide.com, 2019, blog.mide.com/thermodynamic-theory-of-the-ideal-stirling-engine. Accessed 23 Aug. 2023.
[106] Nice, Karim. "How Stirling Engines Work." HowStuffWorks, 4 May 2001, auto.howstuffworks.com/stirling-engine.htm. Accessed 22 May 2023.
[107] Ahmadi, Mohammad Hossein, and Mehdi Mehrpooya. "Investigation of the Effect of Design Parameters on Power Output and Thermal Efficiency of a Stirling Engine by Thermodynamic Analysis." International Journal of Low-Carbon Technologies, vol. 11, no. 2, 1 May 2016, pp. 141–156, academic.oup.com/ijlct/article/11/2/141/2198425, https://doi.org/10.1093/ijlct/ctu030. Accessed 22 May 2023.

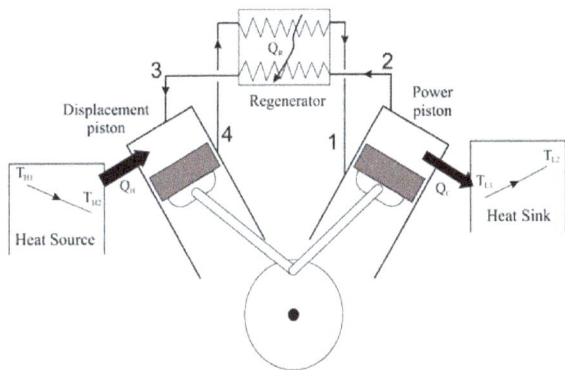

Stirling engines do have drawbacks—most conspicuously, relatively low power output. To maximize efficiency, I would recommend using a variation of a model I've seen sold as a desk toy which uses a total of sixteen pistons (eight hot and eight cold) per engine, connected to a central oblique disk at one end of the driveshaft. Six to eight of these engines could be arranged radially around a central heat source, each extending to an alternator to balance the system load. Another concern is efficiency losses through friction—consider using magnetic repulsion suspension if possible to reduce this. Consider also incorporating a massive flywheel (also to be used as a form of transportation?) to stabilize the momentum of the engine.

Nuclear heat has been used for decades to power "radioisotope thermoelectric generators," which use thermocouples to make "lightweight, compact spacecraft power systems that are extraordinarily reliable" for space travel[108]. But where thermocouples have an average energy efficiency of 3%-8%, heat engines like the Stirling can achieve 30%-50%. With nuclear decay heat as the heat source and cold ocean water as the heat sink (and minimized friction), a huge, multi-cylinder Stirling engine could run smoothly and consistently over a long period and fulfill the power demands of an entire city with infrequent need for intervention. A flywheel or similar mechanism could be used to store and balance rotational energy. Automation from Minerva could reduce the need for maintenance even further by automatically scheduling replacements of parts, performing simple maintenance, and reapplying lubricants when needed.

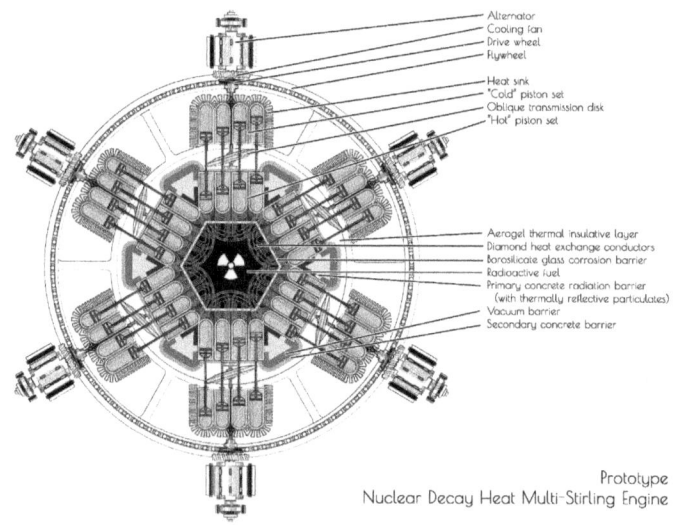

Prototype
Nuclear Decay Heat Multi-Stirling Engine

---

[108] NASA. "Radioisotope Thermoelectric Generators (RTGs) | Cassini." NASA Solar System Exploration, 25 Sept. 2018, solarsystem.nasa.gov/missions/cassini/radioisotope-thermoelectric-generator/ . Accessed 23 Aug. 2023.

Iron Powder

While trying to set fire to a solid iron ingot would be more trouble than it's worth, fine iron powder mixed with air is highly combustible. When this mixture burns, the iron oxidizes and releases energy. While carbon fuel will oxidize into carbon dioxide, iron fuel oxidizes into ferric oxide, which is just rust. Rust can be captured post-combustion and is the only byproduct of the reaction. Iron has a greater energy density than gasoline but is much heavier, so while unsuitable for vehicles or individual homes, it's a (non-radioactive) viable option for renewable energy.[109]

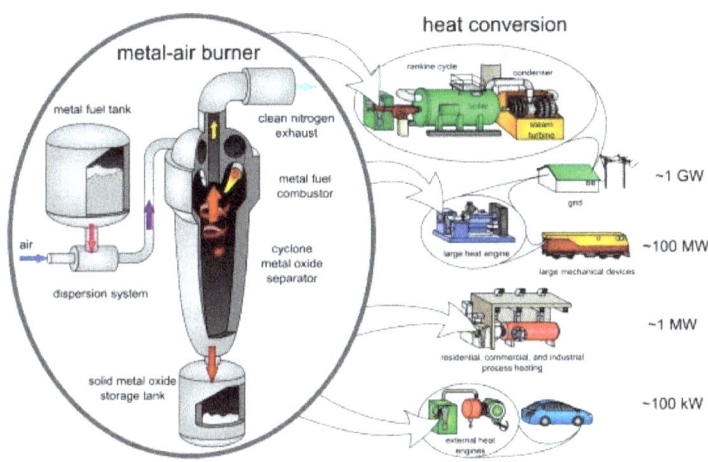

Gravity Battery

With such power sources, on some days the electrical grid will not consume as much energy as it produces. In this case, excess power needs to be stored. I propose the use of gravity batteries as power storage. Gravity batteries operate on a simple concept: a weight is

[109] Ackerman, Evan . "Iron Powder Passes First Industrial Test as Renewable, Carbon Dioxide-Free Fuel." IEEE Spectrum, 13 Nov. 2020, spectrum.ieee.org/iron-powder-passes-first-industrial-test-as-renewable-co2free-fuel. Accessed 22 May 2023.

attached to a pulley system of some height. Excess power is used to pull the weight up into a position of greater potential kinetic energy. When energy needs to be released, the weight is released, and the kinetic motion of turning the pulley is converted back to electrical energy. This process is one of the most efficient means of power storage, mechanically simple, and especially effective in the context of the ocean,[110] as a weight can be lowered down the slope of a continental shelf like an anchor for massive storage potential. This will ensure that Augury has plenty of power in case of an emergency or unexpected load on the system. Excess energy can also be directed into energizing and mineralizing new structures.

XI. Hephaestus Phase 1: Basic Manufacturing Facilities

At first, Augury will have to import everything it needs, including building materials. The conductive framing for the preliminary structure will have to be manufactured on the mainland, then shipped out to the building site, submerged, and assembled. Importing everything we need will get expensive quickly, which is why it's important to establish a manufacturing plant as soon as possible, before the interior construction begins.

Interior structures should be designed with longevity and ease of maintenance in mind. Consider building almost everything from prefabricated, modular, interchangeable parts, such as wall panels, which can be stockpiled and easily swapped out—like the construction equivalent of Legos. Imagine a wall panel already wired for electricity with built-in outlets, removable panels for plumbing fixtures or vents, different versions to include windows or doors. If the panel was damaged or you needed to remodel, you could simply

---

[110] Cazzaniga, R., et al. "DOGES: Deep Ocean Gravitational Energy Storage." Journal of Energy Storage, vol. 14, Dec. 2017, pp. 264–270, https://doi.org/10.1016/j.est.2017.06.008. Accessed 22 May 2023.

unplug it and plug in a new one rather than struggling with running new wire or slow installation times. This may prove to be more a novelty than a practicality, but I believe the concept may have immense potential for efficient construction. Perhaps beyond construction—if we design devices and appliances and whatnot by this same "plug and play" model, parts like motors, power cables, control panels, batteries, or even control boards could be as interchangeable as light bulbs, improving our flexibility in servicing, upgrading, repairing, and inventing machinery. Imagine taking a part out of a kitchen appliance and using it to repair a tool or submarine. The possibilities could be limited only by our creativity.

In general, machines should do the bulk of the work in manufacturing. Humans can do the creative work of designing parts, but whenever possible, repetitive or difficult labor should be programmed by a machine. Programming and artificial intelligence allows us to automate almost anything. Our technology is sufficiently advanced and accessible that no one should have to suffer through any mindless factory jobs. Furthermore, the easier work is and the more we use machines, the more the world of making things (or any kind of work) becomes accessible, even to people who are disabled, elderly, students, and children.

Bear in mind that just as ocean air causes cars in coastal areas to rust faster, inevitable salt in the air in Augury will degrade machinery. Avoid exposed/unprotected steel surfaces as much as possible.

Manufacturing Styles

Many of these styles use computers, which is advantageous—once a design is perfected, machines can automatically manufacture as many iterations as needed from the original file.

3D Printing

Recycled thermoplastics can be repurposed into a number of useful forms, not the least of which is 3D printing filament. Hephaestus should maintain a large collection of 3D printers of varying sizes and

applications. Custom objects and parts can be designed and manufactured onsite as needed for their specific purpose with less waste and lower costs, just as astronauts are able to do in the International Space Station[111]. 3D printing isn't limited to just plastics, either—these days, even entire buildings can be 3D printed with bioresins and concrete[112]. Some high-powered 3D printers are even able to utilize several methods to 3D print with metal.[113] A wide array of printers will allow us to manufacture almost anything we need efficiently and onsite—and with precise 3D computer modeling, we can use specially designed 3-dimensional infill pattern structures which maximize strength and minimize weight[114].

Injection Mold Manufacturing

3D printing is ideal for prototyping—creating few iterations of a designed object—but isn't optimized for mass production. When we have a 3-dimensional object we know we will need to manufacture over and over again, we prototype with 3D modeling and use the prototype to create a mold. Molds can be designed for plastic, but also concrete, metal, and glass—any material which can be liquified. This method (along with a few others) will be useful for creating stock prefabricated construction components.

---

[111] Wall, Mike . "Space Station's Commercial 3D Printer Makes Its 1st Tool (Photos)." Space.com, 14 June 2016, www.space.com/33166-space-station-commercial-3d-printer-first-tool-photos. html. Accessed 19 July 2023.
[112] Bos, Freek, et al. "Additive Manufacturing of Concrete in Construction: Potentials and Challenges of 3D Concrete Printing." Virtual and Physical Prototyping, vol. 11, no. 3, 2 July 2016, pp. 209–225, https://doi.org/10.1080/17452759.2016.1209867.
[113] "Introduction to Metal 3D Printing." Hubs, www.hubs.com/knowledge-base/introduction-metal-3d-printing/. Accessed 3 June 2023.
[114] Podroužek, Jan, et al. "Bio-Inspired 3D Infill Patterns for Additive Manufacturing and Structural Applications." Materials, vol. 12, no. 3, 6 Feb. 2019, p. 499, https://doi.org/10.3390/ma12030499.

CNC Cutters

CNC stands for "Computer numerical control," and refers to machines programmed to maneuver a tool along set paths from computer instructions. CNC cutters refer to a variety of tools: routers, plasma cutters, waterjet cutters, and lasers. These tools can automatically cut flat sheets of anything from paper to wood to stone to metal as specified by a digital design.

Upcycling

In order to reduce, reuse, and recycle waste as much as possible, Hephaestus will maintain a team of workers dedicated to "upcycling." Upcycling, also known as creative reuse, is the process of transforming by-products, waste materials, useless, or unwanted products into new materials or products perceived to be of greater quality, such as artistic value or environmental value. Workers and eventually citizens in Augury will inevitably bring waste into Augury—machines, appliances, tools, and whatever else which eventually wear down and become unusable for their original purpose. Hephaestus workers can use creative problem-solving and critical thinking skills to find value in unwanted items and repurpose them.

In order to make recycling easier, consider stamping all items manufactured in Hephaestus with (color-coded?) symbols to indicate their composition—kinds of plastics, metals, glass, biomass, or minerals. When the items reach the end of their useful life, the symbol will indicate how workers can go about cutting, grinding, or melting them down and combining them with other materials to be reused or composted. These symbols can also apply to public trash receptacles, so citizens know how to sort their garbage.

Whatever materials we produce or resources we harvest must be fully utilized to minimize waste, similarly to how Native Americans would use every part of animals they hunted, rather than just harvesting the meat.

Building Materials

The interior of Augury should be designed with maintenance, longevity, and sustainability in mind. Modern interior architecture is designed to save costs—and sacrifices construction quality. Though using better ingredients will cost more up-front, I believe in the value of investment for the future and delayed gratification. Augury's construction should make use of a wide variety of sustainable materials, including (but not limited to) those listed below.

Recycled Plastic

Plastic is an incredible material, though abused. Why would we use a material that lasts for decades or centuries to make "disposable" packaging, when we could use it to make structures that we *want* to last? In 2017, Kenyan civil engineer Nzambi Matee[115] discovered a way to convert plastic waste into building materials by heating it, mixing it with sand, then compressing it. These blocks can be formed to any size and are strong enough to be used for wall construction or paving. Plastic is also commonly recycled into a material known as plastic lumber, which is as strong as wood but easier to work with, and immune to water, insects, and rot. Many kinds of plastic can even be recycled into 3D printer filament, allowing a nearly infinite variety of applications.

Collecting and recycling plastic waste from the nearby North Atlantic Garbage Patch would benefit the ocean ecosystem and provide a consistent supply of free building materials.

Bioplastics

Petroleum-based plastics have a number of useful properties, most notable of which is their longevity—they last a very long time, but not forever. Their capacity to be recycled degrades with each use, and as they are petroleum based, they are inherently a finite resource

---

[115] UNEP. "Nzambi Matee." Young Champions of the Earth - UN Environment Program, www.unep.org/youngchampions/bio/2020/africa/nzambi-matee. Accessed 2 June 2023.

(not to mention the toxic costs of production). "According to the U.S. Energy Information Administration's International Energy Outlook 2023, the global supply of crude oil, other liquid hydrocarbons, and biofuels is [only] expected to be adequate to meet the world's demand for liquid fuels through 2050 [and] there is substantial uncertainty about the levels of future liquid fuels supply and demand.[116]" Eventually we will need to turn to a more renewable version, in the form of bioplastics. Bioplastics either originate from a renewable resource, are biodegradable, or are both.[117] They can be made from biomass materials like vegetable fats and oils, cellulose, corn starch, straw, woodchips, sawdust, recycled food waste, and have very similar properties to petroleum plastics.

Recycled Textile Bricks

There's no question that waste accumulation is becoming a big problem—and with the advent of fast fashion, textiles are a huge contributor. French research and development firm FaBRICK has found a way to use recycled clothing waste as a tensile reinforcer when combined with a binding agent to make bricks. The resulting material "has a very good mechanical resistance almost like a concrete block...has a good insulating quality, acoustic and thermal, and a good resistance to fire. Hence [they] can make panels and tiles that improve the ambiance of a room or a public space. It also has a good water resistance,[118]" at least, enough for interior use.

---

[116] "Does the World Have Enough Oil to Meet Our Future Needs? - FAQ - U.S. Energy Information Administration (EIA)." Eia.gov, 2016, www.eia.gov/tools/faqs/faq.php?id=38&t=6.
[117] Ashter, Syed Ali . "Bioplastics - an Overview | ScienceDirect Topics." www.sciencedirect.com, 2016, www.sciencedirect.com/topics/engineering/bioplastics. Accessed 20 July 2023.
[118] Soulié, Pierre-Louis . "FabBRICK: Construction Materials from Recycled Textile." DesignWanted, 16 May 2021, designwanted.com/fabbrick-construction-materials-recycled-textile/. Accessed 3 June 2023.

Concrete Impregnated Fabric & Shotcrete

Concrete Cloth or **Concrete Impregnated Fabric** (CIF) is a flexible, concrete filled geosynthetic which provides a thin and durable concrete layer when hydrated. Its typical use is as an erosion/pest/weed barrier, but can be used to quickly manufacture structures around a simple frame like a tent. Before hydration, the material is flexible and can easily be cut to any shape necessary, as well as stitched to other pieces to form more complex shapes. CIF pairs well with geodesic domes to form efficient and easily assembled structures.

Additionally, **Shotcrete** is a method of applying concrete projected at high velocity onto almost any surface, even vertical or overhead; the impact created by the application helps the concrete to set up. This method allows concrete construction in even very tight spaces and "free form" applications. Shotcrete also has higher compressive strengths than cast-in-place concrete and lower construction costs. Shotcrete would be an excellent option for patching breaches while the exterior wall remineralizes.

Autoclaved Aerated Concrete (AAC)

Concrete alone is not a particularly good thermal insulator—but mixing liquid concrete with a foaming or expansion agent (like a detergent) fills it with air bubbles. The resulting material is called **Autoclaved Aerated Concrete**, which is about 80% air by volume. Compared to typical concrete, it is lighter, more insulative (both for heat and sound), easier to work with (drilling holes and cutting), resistant to water, decay, and insects, fire resistant, less dense, and has a higher thermal mass. AAC can be formed into both blocks and panels for versatile applications. As the mix is made in liquid concrete mix, AAC can also be used for shotcrete and 3D concrete printing to produce entire buildings in only a few days.

Cob & Adobe

Cob and adobe, put simply, are dirt. Dirt is one of the oldest and long-lasting building materials, with some examples over a thousand

years old. After mixing a blend of clay, sand, straw, and water, cob can be molded into a monolithic structure—but after it dries, it's as strong as concrete. It also has an excellent thermal mass, is breathable, fire-resistant, and is an excellent insulator.[119] It's also easy to work with and repair, though its weight could make the cost of transportation not worth the effort.

Glass

Glass is an excellent material—infinitely recyclable, immensely versatile, and nearly chemically inert.[120] Wherever possible, glass should replace plastic in Augury as a packaging material. But glass's potential doesn't stop there. Glass or fiberglass can be used to make containers, insulation, gravel, tile, composites, and construction materials—including, of course, interior windows. When more glass is needed, Hephaestus can manufacture it from high-silica ocean sand.

Renewable Materials

Many quick-growing plants can be used as building materials. **Bamboo** grows quickly and abundantly all over the world, considered a weed in the southeastern United States. As it is in the grass family, it grows flexibly and can be formed into useful shapes like arches and cross-sections. With proper treatment, bamboo can be used to build strong framing, supports, walls, ceilings, and more—especially when combined with other materials. With higher tensile strength than steel, good fire resistance, elasticity, low weight, and low cost, bamboo is versatile enough to compose an entire building; in fact, the Green Village in Bali is made almost entirely of bamboo.[121] It's not without its drawbacks, of course—bamboo

---

[119] Koru Architects. "Benefits of Building with Cob, plus Examples." Koru, 19 Sept. 2017, koruarchitects.co.uk/cob-super-natural-materials. Accessed 3 June 2023.
[120] Glass Packaging Institute. "Glass Recycling Facts - Glass Packaging Institute." Gpi.org, 2020, www.gpi.org/glass-recycling-facts. Accessed 3 June 2023.
[121] Beser, Ari. "In Bali, Bamboo Architecture Offers Model for a Sustainable Future." National Geographic Society Newsroom, 7 Feb. 2016,

requires preservation (and lots of water, depending on variety), has a tendency to shrink, and can be weak to insect and fungal damage.[122]

Straw grows quickly and easily—it takes more effort to stop it from growing than it does to let it take over a field. Despite the fairy tale, when compressed into bales, straw is a strong building material and an excellent insulator. When combined with lime and a binder, **hemp** forms a promising material known as **hempcrete**–the woody hemp absorbs and stores a significant amount of carbon, and provides tensile strength and greater insulative properties as a building material. Hemp can also be used in the manufacture of plasters, insulation wool, and subflooring mats. **Seaweed, seagrass,** and **algae** provide natural resources, and can be grown locally. **Cactus** can be used to make a leather alternative. **Mushrooms** can also be used to produce a leather alternative, along with building blocks stronger than concrete.[123] Some of these materials would require too much space to be grown locally and would have to be imported.

Self-Healing Materials

Some special materials will regeneratively recover from flaws over time, reducing maintenance costs and increasing the longevity of the structure. Concrete has a few self-healing versions—some chemical, some mineral, and some bacterial.[124] Self-healing qualities are especially useful for concrete, to compensate for its brittleness. Self-healing doesn't stop at concrete, however—MIT has been

---

blog.nationalgeographic.org/2016/02/07/in-bali-bamboo-architecture-offers-model-for-a-sustainable-future/. Accessed 20 July 2023.
[122] The Constructor. "Bamboo as a Building Material - Its Uses and Advantages in Construction." The Constructor, 5 Dec. 2016, theconstructor.org/building/bamboo-as-a-building-material-uses-advantages/14838/. Accessed 3 June 2023.
[123] Mershin, Andreas. "These Mushroom Bricks Are up to Twice as Strong as Concrete." MIT Media Lab, 26 July 2022, www.media.mit.edu/articles/these-mushroom-bricks-are-up-to-twice-as-strong-as-concrete/. Accessed 17 July 2023.
[124] Multiple authors. "Self Healing Concrete - an Overview | ScienceDirect Topics." Www.sciencedirect.com, www.sciencedirect.com/topics/engineering/self-healing-concrete. Accessed 2 June 2023.

working on a gel that grows, strengthens, and repairs itself using carbon dioxide.[125]

Untreated Materials

Note that the rules for construction for some parts of Augury's interior will be different than on the surface—structures and materials are not exposed to extreme temperature fluctuations, precipitation, or direct sunlight, which opens the door in some cases to use of less durable materials like untreated wood, plastics, fabric, and so on. Lower exposure to moisture will significantly reduce the risk of rot (except in atriums where the ground and air are kept moist for plant health). Absorbent materials will still soak up moisture from inevitable spills and splashes and should be used sparingly or protected with a thin layer of wax or oil, but will not need to be treated with toxic chemicals, such as those used to pressure-treat wood.

XII. Hestia Phase 3: Interior Architecture

Inside the mineral aggregate shell, Hestia workers will be free to construct bespoke structures as needed. Support beams can be embedded directly into the exterior stone, so interior walls, ceilings, and floors can be made of relatively weaker materials. Importing construction materials from the mainland will be expensive, so using locally sourced materials will be preferable. With a reliable local supply of prefabricated parts from Hephaestus, Hestia should be able to furnish interior structures and spaces with ease. Utilities (pipes, wiring, vents, etc.) should either be mounted inside the walls where they can be easily accessed, or behind wall panels with enough space to make maintenance and repair convenient. The vast majority of interior architecture in Augury should be made using stock prefabricated parts (manufactured by Hephaestus), such as

---

[125] Biggs, John. "Self-Repairing Construction Materials Could Have Huge Benefits." Jobsite by Procore, 21 Mar. 2022, www.procore.com/jobsite/self-repairing-construction-materials-could-have-huge-benefits. Accessed 17 July 2023.

wall panels, windows, and building blocks all designed to fit together easily with little effort or design, so as to make repairs and replacement easy.

A full team of architects and engineers will be needed to properly design the interior of Augury, but in the meantime, I propose the following design as a conceptual starting point:

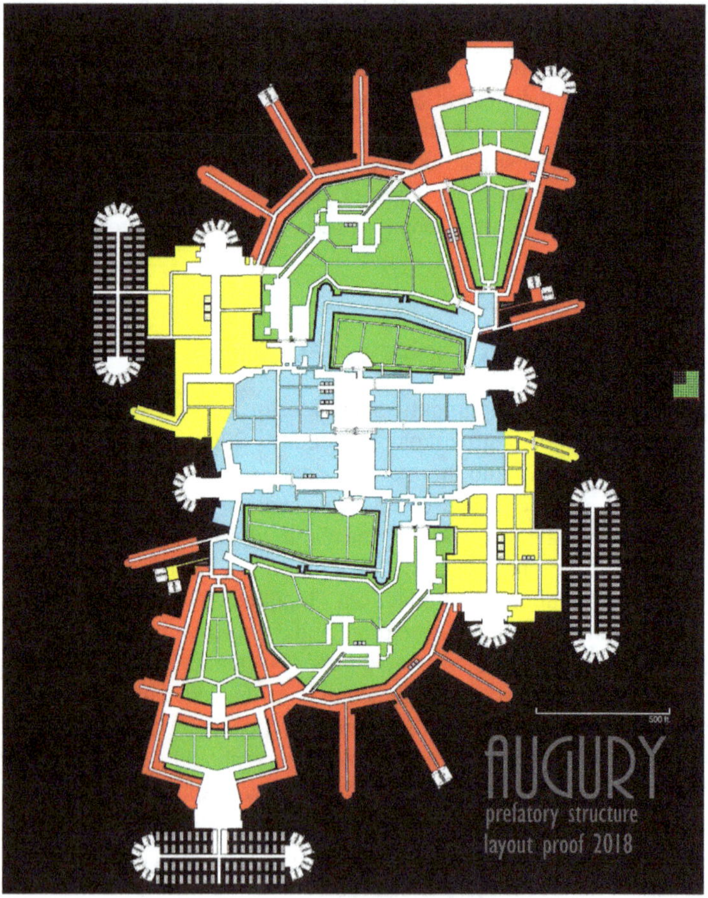

Note that the structure is widespread and sprawling, rather than clustered under a central dome. Though expansive domes are a common feature for underwater habitations in science fiction, they bring a risk of critical failure and should not be used to encapsulate large areas of the city. Augury should be constructed in sectors, each of which can be sealed off at bulkheads in case of a breach. To reduce pressure variation, the overall structure of Augury should be sprawling, covering horizontal space rather than rising very high. Note also that sections are color-coded (though this will be difficult to perceive if your version of this manuscript is in black and white). Office space and municipal facilities (blue) are clustered at the center, with commercial and industrial facilities (yellow) on either side (connecting to the side submarine/airlock bays). On the top and bottom are thin residential areas (red) wrapped around wide open atriums, which will primarily be green spaces. Residential areas are narrowly designed to give either views of the ocean or an atrium, to reduce claustrophobia. Main concourses are in white.

One critical feature this design lacks, however, is an absence of sharp corners— round shapes are far better at withstanding pressure than corners, so all the outermost walls should be curved and smooth (if the internal pressure is equivalent to the external pressure, however, this will no longer be an issue).

My blueprint may come with other issues I haven't noticed—I would defer to a professional firm of architects and engineers for the final design. Overall, however, Augury should be designed in the style of large buildings such as hotels, airports, and shopping malls, which are made to make an enclosed structure feel open and airy (which will reduce claustrophobia). Shown on the following page is a photo (courtesy of The Arcade Providence) of a former shopping mall converted to apartments that reflects this intended style.

Consider the layout style of an indoor shopping mall from the example below of the Post Oak mall in College Station, Texas: wide walkways (made for foot traffic, but usable by small vehicles if needed) with compartmentalized spaces on the sides, easily navigable, high ceilings, usually including food courts and entertainment facilities, and ample sitting space. These buildings are designed to keep you comfortably inside as long as possible; as such, they are perfect for our purposes.

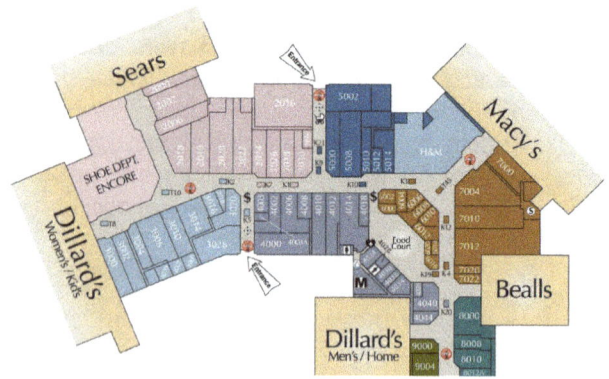

The final example below represents the ideal style for interior structure in Augury: the Gaylord Opryland Resort & Convention Center in Nashville, Tennessee, which uses multiple cavernous atriums to enclose sprawling green spaces and even smaller buildings on a series of islands, bordered by hotel rooms overlooking artificial ponds and rivers. Features like these will minimize discomfort in an underwater facility. Natural light would stream in from the surface through skylights, and ample plant life will improve general mental health and air quality.

If funding will not allow the construction of such a large structure all at once, the structure can be built in stages according to this building plan. Proposed facility layouts reflecting these stages are displayed in illustrations.

The first facilities to be fully developed within Augury should be a fully-outfitted hotel, equipped with all the facilities included on a cruise ship (which, like Augury will be at first, is essentially a hotel in the middle of the ocean). This will include facilities such as an exercise room, medical facilities and morgue, a dining area, a pool, movie theater, entertainment facilities, laundry facilities, and so on.

As soon as these facilities are built (and the overall structure is confirmed to be safe), we can start accepting guests and building national revenue. If funding is tight, the living quarters for guests can at first be constructed in the style of "glamping." I imagine many people would pay well for the opportunity to camp in the world's first underwater city as it is being built. Tours exploring the construction progress could be offered as a perk or additional revenue stream.

Everywhere in Augury intended to be accessed by the general public **must be wheelchair accessible**, without exception (except for maintenance areas).

In no case should the faux rock style of architecture (which typically uses shotcrete sprayed on a wire form, commonly used in zoos and amusement parks) be used to give the illusion of natural rock in atriums. This looks cheap (however, more advanced techniques for sculpting concrete and plaster can produce a convincing and pleasing stone effect). Likewise, walls and ceilings should not be painted to "give the illusion" of extending towards a forest or field or the open sky. Murals are one thing, but pretending an enclosed room is bigger than reality is quite another. Everyone in Augury will know they are indoors; there is no need to pretend otherwise. We should embrace walls and ceilings where they are and make them beautiful through good architecture, not pretend they aren't there. As an alternative, use stonework styles designed to resemble ancient temples and civilizations for plant-heavy areas, especially around water features. This style ages well and can handle the weathering.

Use of Art Deco

In the early 20th century, the Art Deco movement combined modern styles with fine craftsmanship and rich materials, and represented luxury, exuberance, and faith in social and technological progress.[126] I think we benefit from these sentiments in our culture, and propose we make extensive use of the Art Deco (and recent revival, Neo

[126] Wikipedia Contributors. "Art Deco." Wikipedia, Wikimedia Foundation, 28 July 2019, en.wikipedia.org/wiki/Art_Deco. Accessed 6 July 2023.

Deco and related style, Streamline Moderne) style in the design of Augury. The dramatic Art/Neo Deco style is characterized by use of jewel tones juxtaposed with metallic colors (usually gold and brass), use of glass next to polished metals and leather, amber lighting, dramatic vignettes, long horizontal lines, smooth and curving surfaces, gentle curves next to crisp corners, and occasionally nautical elements to represent the Transatlantic culture. Aside from being visually beautiful, the style is timeless, ages well, translates into a great variety of mediums, and the strong lines and use of geometry pairs well with industrial arts, which serves to celebrate the utility machinery we rely on rather than hiding it in closets or in the ceiling. The style applied to not just architecture and furniture, but to machinery and tools as well.

Of course, though I believe Art Deco should be Augury's primary interior style, it should not be exclusive—there should be neighborhoods and districts which showcase other beautiful styles, like Victorian, Tudor, Gothic, and so on—perhaps consider stoneworking in the style of the ancient civilizations of the world, which age beautiful and take to moss and other plants. Some aspects of brutalism (especially eco-brutalism) even have their place, as the style emphasizes the engineering design of a building and showcases the bare building materials and structural elements. Look to American shopping malls in the 1970s to 1980s; these spaces were designed to comfortably enclose large groups of people and high traffic with multipurpose development. It's important also that whatever style we use, we use it intentionally, and for the sake of our mental health consider building as a medium of art in architecture and interior design, rather than simply out of utility. The styles we use should have visual appeal and personality. Avoid neutral tones and brutalist design styles. Our surroundings are an expression of who we are and how we see ourselves. The spirit of the Art Deco movement is in the name, which is short for "arts decoratif," or "decorative arts." By decorating and designing decoratively rather than purely for practicality, we make the world around us more beautiful. Let's surround ourselves with intentional beauty, even in the mundane. People care more about the space they live in when even ordinary things like benches, facades, and mailboxes have

artistry in their design. This is a form of public trust and an investment into the community.

Furnishings

Insisting upon matching furniture and furnishings throughout Augury would be prohibitively expensive and unnecessary. How would it serve us, to present a unified front of conformity? Instead, we should have minds for upcycling, making use of recycled furniture, repairing and recycling it when possible, not vainly obsessing over coordination and matching with interior design. We can also combine form with function and pursue modular furniture (consider for example the work of designer Joe Colombo). Augury's interior spaces should look organic, lived in, not flawless and sterile. Every corner of Augury should be distinct, unique, with its own personality. Avoid use of colors in the taupe family; life is too short for neutral tones. Interior design in Augury should avoid the modern minimalist neutral style that is so tragically popular right now.

XIII. Apollo Phase 1: Local Network

In our modern age, a home internet connection should be considered a basic utility: Like indoor plumbing and electricity, not technically a necessity for survival, but so crucial for modern life as to be considered a given by employers and schools. Internet service in America is a mess, largely due to an oligopoly held by the largest service provider corporations[127][128], and those same companies have lobbied the federal government to make it illegal for cities to build

---

[127] Press, William H. "Monopoly/Oligopoly - Status, Behaviors, Etc. Of Other Industries." WPressUTexas.net, University of Texas at Austin, 23 Feb. 2015, wpressutexas.net/cs378h/index.php?title=Monopoly/oligopoly_-_status,_beha viors,_etc._of_other_industries. Accessed 20 July 2023.
[128] Moore, Heidi. "Price-Gouging Cable Companies Are Our Latter-Day Robber Barons | Heidi Moore." The Guardian, 4 June 2013, www.theguardian.com/commentisfree/2013/jun/04/price-gouging-cable-comp anies. Accessed 20 July 2023.

their own network[129]. Service is unreliable, slow, and spotty, and on top of all that, expensive—comparatively far worse than other industrialized nations. Building your own internet service, however, is either illegal or impossible–so we're trapped with limited options. We also enjoy very little privacy online, as numerous occasions have shown to us that our online presence is constantly recorded and used to manipulate[130] or surveil us.[131] A lot of people value their privacy very highly, while others are unconcerned—but I believe with such an internet-based international industry and economy, we should have as easy and inexpensive access to reliable, strong internet as we do electricity and water, without concern over how the provider might use the service against us or sell our data.

In Augury, direct access to undersea communications cables will allow the Apollo department to hardwire fiber-optic ethernet connections all over the city. Bermuda, for example, is connected to five undersea cables: GlobeNet, Gemini Bermuda, Caribbean Bermuda U.S., and Challenger Bermuda 1[132]. This direct connection will be open-access to all citizens. Any citizen will be able to visit the Apollo department section of the Augury city website for metrics on speed, bandwidth, cost, and so on. Cost of maintaining the internet connection will be taken from the national income, at no additional charge to all citizens. Unfortunately, it may be unwise to offer unlimited data as some individuals might abuse this and transmit enormous quantities of data—if there is a monthly data limit, it should

[129] Brodkin, Jon. "ISP Lobby Has Already Won Limits on Public Broadband in 20 States." Ars Technica, 12 Feb. 2014, arstechnica.com/tech-policy/2014/02/isp-lobby-has-already-won-limits-on-public-broadband-in-20-states/. Accessed 20 July 2023.
[130] Manjoo, Farhad. "Facebook Followed You to the Supermarket." Slate, 20 Mar. 2013, www.slate.com/articles/technology/technology/2013/03/facebook_advertisement_studies_their_ads_are_more_like_tv_ads_than_google.html. Accessed 20 July 2023.
[131] Empson, Rip. "Google Biz Chief: Over 10M Websites Now Using Google Analytics." TechCrunch, 12 Apr. 2012, techcrunch.com/2012/04/12/google-analytics-officially-at-10m/. Accessed 20 July 2023.
[132] TeleGeography. "Submarine Cable Map." Submarinecablemap.com/, 2019, www.submarinecablemap.com/. Accessed 20 July 2023.

be so high that the average user never even approaches it. 800 GB of data per month is a common limit from cell providers. Visitors will have access to a separate network, so that heavy tourism does not interfere with citizens' online activities.

Pornography websites should be blocked, for reasons outlined in Part 3: Mental, Emotional, & Social Health: Addiction.

Radio EMF Concerns

In the enclosed environment of Augury, radiation from low frequency (like radio waves) electromagnetic fields (EMF) from devices like power transmission lines and wireless networks may be of greater concern. Note that this radiation is low-**frequency** but not low **energy**. On the surface, this energy can radiate out into the air and dissipate, but with Augury's thick stone walls, we are more likely to be exposed. "The World Health Organization…classifies extremely low frequency electromagnetic fields as *possibly* carcinogenic to humans based on limited evidence…however, scientific studies have **not consistently shown** whether exposure to any source of EMF increases cancer risk[133]." Furthermore, though research data is similarly limited, "honeybees are sensitive to pulsed electromagnetic fields generated by [wireless technology," which can temporarily disorient them[134][135]; this phenomenon may also affect other animals like birds and fish which possess magnetoreception. In an environment saturated with radiant energy, honey bees might be stressed and dissuaded from establishing hives or producing honey.

---

[133] US EPA,OAR. "Electric and Magnetic Fields from Power Lines | US EPA." US EPA, 8 July 2019, www.epa.gov/radtown/electric-and-magnetic-fields-power-lines. Accessed 4 Aug. 2023.

[134] "Cell Phone Bee Mortality Link: Sensationalism Not Science." Vanderbilt University, 14 June 2011, news.vanderbilt.edu/2011/06/14/cell-phone-bee-mortality-link-sensationalism-not-science/. Accessed 4 Aug. 2023.

[135] Jungwirth, Joseph, "The Effect of Electromagnetic Fields Produced by WiFi Routers on the Magnetite (Fe3O4) Particles of the Italian Honey Bee (Apis mellifera ssp. ligustica)" (2019). Culminating Projects in Biology. 42. https://repository.stcloudstate.edu/biol_etds/42

The best way to mitigate these risks may require further research, but wired data technology should be preferred in Augury and wifi transmitters should be minimized in the atriums. If analog/lower power radio signals do not have the same negative effects as digital, handheld transceivers (walkie talkies) or something similar can be used for short range communication. Though it is an extreme measure, protecting atriums in a Faraday cage would protect them from overwhelming EMFs. **However**: while this issue should be monitored, it is highly unlikely to warrant any serious action. Bees are adaptable, and given time and generations, will adjust to the unusual environment. Power transmission lines should be safe if properly shielded and insulated, and should also be outfitted with ground fault circuit (GFC) sensors, as current leakage into ocean water is a concern.

XIV. Apollo Phase 2: Data Center

As connectivity in Augury increases, it's time to start pursuing the primary source of national income for Augury: submerged data centers. The digital real estate industry grows more and more every year, and one of the greatest costs of maintaining servers is cooling. Ocean water is very cold and able to absorb a lot of heat (due to water's unusually high heat capacity[136])—and as companies like Microsoft have already found, "underwater data centers are reliable, practical, and use energy sustainably."[137] With direct access to undersea communications cables, near limitless energy, and an abundance of cold water, Augury will be able to generate most of its national income from renting server space in underwater data

---

[136] Water Science School. "Specific Heat Capacity and Water | U.S. Geological Survey." Www.usgs.gov, 6 June 2018, www.usgs.gov/special-topics/water-science-school/science/specific-heat-capacity-and-water. Accessed 20 July 2023.
[137] Veda. "Stepping into the Era of Underwater Data Centers! But Will They Work?" Analytics Insight, 16 Sept. 2022, www.analyticsinsight.net/stepping-into-the-era-of-underwater-data-centers-but-will-they-work/. Accessed 20 July 2023.

centers, run by Apollo. It's a stable, scalable source of income that reliably grows as online industries increase. Affordable, high-volume data storage will be a valuable resource to even very large corporations worldwide, and if facilities are efficiently maintained and well-constructed, generate passive income. Having Augury built around this industry will doubtlessly attract companies specializing in technology, research & development, software development, and video game development (which will be further explored in Part 3: Economy: Video Game Industry) and give Augury a stable advantageous position in the modern economy and promote the use and improvement of technology within the city culture. This will also create a higher demand for technology manufacturing from Hephaestus.

Servers can be built to be completely submerged in saltwater, to maximize the cooling effect. If this proves practical (which it may not, due to biofouling, in which case it can be immersed in a liquid like mineral oil), the entire data center can be submerged and serviced by divers. The absorbed heat could possibly even be diverted to heating some other part of the city. As technology improves, data centers may even be able to be operated remotely, so they can be situated in much deeper (and colder) water, further increasing profit margin. Eventually, the in-house supply of high volume data storage will naturally facilitate projects related to virtual reality, artificial intelligence, supercomputers for research and simulation, and more.

Knowledge collected by Apollo should be (with very few exceptions) freely shared with the rest of the world; not hoarded.

XV. Aegle Phase 2: Full Medical Facilities

At this stage, Augury should be a self-sustaining habitable environment, even able to fund itself. The first application of this funding should be towards the upgrading of our central medical facilities. From this point forwards, no denizen of Augury should ever

have to worry about receiving adequate healthcare. Healthcare must be publicly run, publicly funded, and publicly offered through the Aegle department—the physical wellbeing of the citizens should never be operated for profit or without oversight of said citizens. Population and tourism intake should be closely monitored at this stage to ensure that medical staff are not overwhelmed or overworked. With a little help from Minerva, Aegle will eventually be able to process higher quantities of data to provide prognostics, diagnostics, and treatment more efficiently. To name a few examples, machines can help Aegle workers perform and analyze noninvasive scans, design and 3D print casts for broken bones, analyze blood, operate surgical robots, sequence genes, and recommend medicine based on biological data. At any point, the human technician will have the ability to override or disable the machine assist if it becomes a hindrance.

This first central healthcare facility should be built with a very high capacity, at least 10%-20% higher than the total population at all times to prevent the possibility of overflow. Space should also be allocated for future expansions. Empty multipurpose/storage space could also be maintained in case of internal or external refugees. Furthermore, the spread of disease is a major concern in an enclosed, air-tight space like Augury. Fortunately, there's a simple way to help reduce it: bacteria, yeasts, and viruses are rapidly killed on metallic copper surfaces due to the oligodynamic effect,[138] so Aegle facilities (as well as other high-traffic areas) should be fitted with copper or brass fixtures, including handrails, light switches, and more. This paired with nanofilters and UV lighting in healthcare facilities will greatly reduce spread of pathogens.

The facility should have a central location that is easily reached, with dedicated infrastructure established by Hermes to minimize the time before a patient can receive treatment. Consider installing some kind of automated emergency transport system that fulfills the same need

---

[138] Charles F. McKhann, Harve J. Carlson, Harriet Douglas; Oligodynamic action of metallic elements and of metal alloys on certain bacteria and viruses: I. In Vitro Observations. Pediatrics September 1948; 2 (3): 272–289. 10.1542/peds.2.3.272

as ambulances—perhaps some pneumatic capsule network, or a dedicated high-speed tram system. In such a dangerous environment as the ocean, Aegle and its facilities should maintain emergency contingency plans for every eventuality and drill them regularly. Preventative maintenance will make a big difference to this end—Take a potential pandemic, for example. In a closed system like Augury, there's no room for optional protection against deadly disease. If a disease becomes a significant issue in Augury, Physicality and Aegle will work closely together to develop a vaccine or other treatment, or acquire one from another country if necessary. This treatment would become mandatory to all eligible citizens, to protect the health of citizens too weak or otherwise unable to accept the treatment. Electing to abstain from the treatment, which should be considered dangerously selfish, should result in expulsion or revocation of citizenship.

Beyond emergency preparedness, Aegle will also pursue the improvement of the quality of life of humanity. This includes facilities for manufacturing medication, so as to provide them at minimal cost. Insulin, for example, can be easily manufactured using genetically modified bacteria—and since people need it to live, shouldn't it be inexpensive? Much like the Open Insulin Foundation,[139] Aegle should have facilities to manufacture insulin, providing it for free to citizens and eventually distributing it internationally at cost, as well as working with Hephaestus to manufacture the equipment so people can manufacture insulin themselves. Additionally, Aegle should also house facilities for the design and manufacture of a wide variety of prosthetics and replacement parts. Modern science has progressed to the point of being able to manufacture excellent replacements for a variety of body parts, including bones, internal organs, skin, and entire limbs. Aegle should be constantly researching and improving these methods to help replace what victims of serious injuries have lost, and make their lives easier. Sometimes, however, advanced technology isn't the answer to an ailment, and the simplest solution

---

[139] "Open Insulin Project." Open Insulin Project, openinsulin.org/. Accessed 20 July 2023.

is often the best. Aegle workers should avoid polypharmacy[140], the overuse of prescribed medication, and understand that prescribing medication is sometimes unnecessary when a change in lifestyle will more effectively alleviate the problem. The flexibility of lifestyle afforded by Physicality and Mentality enables the individual to pursue simple solutions for simple problems, like tea for a stomach ache or pineapple juice for a cough—but encourages them to seek professional medical help for a persisting ailment.

Aegle should also pursue the possibilities of gene therapy—to repair, *never* to improve (on humans or animals). CRISPR Cas9[141] is a promising new technology which, simply put, enables researchers to "cut and paste" genes very easily and inexpensively. Geneticists have just begun to explore the potential of this inexpensive and easy gene therapy, so Aegle facilities should oversee a research lab dedicated to potentially beneficial applications. These applications may include anything from curing genetic conditions and maladies to perhaps even some kinds of cancer. Eventually, technology provided by Apollo and Minerva will enable Aegle to run powerful simulation programs for gene sequencing to improve research. Demeter can also use CRISPR to produce more durable strains of plants which can thrive with reduced space and sunlight more easily, or even underwater.

To summarize: the main medical facilities of Augury should serve as a care and therapy ward, but also as a top-of-the-line research center. The goal of Aegle should be the improved physical health of all people—citizens prioritizes, but not citizens alone. It should share data and resources worldwide as it is able to. Healthcare is not a competition. It is a common goal.

---

[140] Katella, Kathy. "Do You Really Need All of Those Medications?" Yale Medicine, 31 Aug. 2020, www.yalemedicine.org/news/polypharmacy. Accessed 4 Aug. 2023.
[141] MedlinePlus. "What Are Genome Editing and CRISPR-Cas9?" Medlineplus.gov, Medlineplus, 22 Mar. 2022, medlineplus.gov/genetics/understanding/genomicresearch/genomeediting/. Accessed 20 July 2023.

XVI. Hestia Phase 4: Finishing Work

At this stage, Hestia can begin the finishing stages of construction. This includes facing, plastering, woodworking, wallpapering and glazing, the installation of most interior walls and windows, insulation, finishing wall panels that can be painted, plumbing and electrical/lighting fixtures, flooring and carpeting, furniture, and so on. This stage also includes infrastructure for later phases such as support structuring for Hermes transport networks and building up landscaping structures for Demeter. Hestia will also complete the hotel facilities at this stage, so that Augury will now be fully equipped to receive guests. If possible, consider building a convention center, as this would be another significant national revenue stream.

After the hotel is completed, permanent residents can begin temporarily inhabiting the rooms while residential sectors are being constructed. Apartments in Augury should be comfortably sized—not small or cramped, but modest enough to use space efficiently. Cabinetry should span floor-to-ceiling to make the most effective use of limited space, and kitchen appliances should be multi-function. Walls are thick enough to block out all the sound from neighbors and equipped with effective and locally controlled HVAC units.

Remember: everywhere in Augury intended to be accessed by the general public **must be wheelchair accessible**, without exception beyond maintenance areas.

XVII. Hermes Phase 2: Transportation Network

As Augury grows in size, it's important that people and materials can get where they need to go efficiently. Augury should be easy and intuitive to navigate. Consider implementing, as some cities do, colored lines painted into the pavement along major paths, which pedestrians can follow to landmarks. In an enclosed environment

where people cannot cross fields and roads to find the shortest path between two points, proper city planning is vital to ensure optimum convenience. Concourses, elevators, stairs, escalators, ramps, and so on should allow pedestrians to follow the shortest and most direct paths to their destinations. If a public tram system of some kind circling the city has not been constructed by this point, it should now.

Hermes should maintain headquarters at the lower opening of the main intake shaft, from which materials and people are transported down from the surface. From there, they can at this point start to expand their facilities to an industrial allocation center of omnidirectional conveyor belts for efficiently distributing cargo all over the city. Hermes will be in charge of managing and documenting the comings and goings of people, exports, imports, and vehicles that make contact with Augury, so their facilities will resemble an airport of sorts (and all the associated facilities). Technology can streamline and automate this process.

If financially viable, Hermes can consider at this point collaborating with Hephaestus to start manufacturing vehicles in-house; manned, unmanned, marine, submarine, and aerial. Eventually Hermes should maintain a fleet capable of servicing the entire city, as well as lifeboats and escape pods.

Relocation Services

Moving to Augury won't be easy. Hermes should manage an agency dedicated to assisting new citizens in relocation—this will include everything to transportation of the individual and their family, their belongings, and filing the necessary paperwork.

XVIII. Demeter Phase 2: Interior Ecosystem

In order to be self-sufficient (and maintain a healthy environment for humans), Augury must maintain a balanced natural interior environment. This includes plants of every shape and size arranged in a permaculture "food forest" arrangement: the canopy, sub-canopy, shrubs, herbaceous, ground cover, root crop, and climbing vine layers. Flowering plants will also be needed to attract pollinators and increase natural beauty. Invasive plants should be avoided or carefully contained. Grasses should not be included at all; consider moss or creeping red thyme as a ground cover plant. However, plants alone are not enough—a complete ecosystem also includes animals. Though Augury won't have room for large livestock animals like cows, pigs, or horses, smaller livestock like miniature pigs, rabbits, sheep and goats may be sustainable. Poultry like chicken, duck, and goose are viable options, and benefit many systems due to their diet and activities. Aquatic animals like fish, shellfish, and molluscs are extremely promising. Many insects are beneficial to ecosystems as pollinators and more, and are excellent sources of protein. Fungi play an important role in soil health, are important decomposers, and are an excellent food source. Even bacteria play a crucial role in how they affect the health of the soil and water. Each organism offers a unique combination of advantages and drawbacks, each of which fit together in different ways. Demeter's job is to balance the combination of organisms into a well-functioning system. Life does not (and should not be forced to) work in isolation. The ecosystem is a complex machine, and the most sophisticated technology on the face of the earth. To make the interior of Augury work alongside it, to use a biophilic design rather than work against it will greatly benefit us and reduce our energy waste.

Food

Food in Augury will look different than it usually does on the surface. It should—the typical modern diet in industrialized countries is extremely unhealthy. This is largely due to the industrialization of food, which is something that will be reduced in Augury. In such a small system, most food will be "farm-to-table" in an extremely short amount of time. Food will be minimally processed, grown and harvested locally, and have fewer artificial ingredients. Food comes exclusively from other life forms, and with the cost of importing, the bulk of our caloric intake is going to have to be what we grow and harvest ourselves. We're accustomed to space-intensive foods like beef, wheat, corn, and rice, which we won't have room for. We have the wide expanses of the ocean floor to propagate plants like kelp and seaweeds as sources of food and natural resources (and for farming fish and shellfish), and it's conceivable that eventually we could maintain floating barges to grow small amounts of these field crops—but not enough to sustain ourselves or make the effort worthwhile. We're going to have to start thinking more efficiently, more practically, less luxuriously.

There are a variety of space efficient caloric and nutrient dense foods available. Fruiting shrubs, vines, and trees in general are more sustainable as they don't have to be replanted every year, and without risk of frost, will grow bigger and stronger over time. Some fruit trees can have multiple varieties grafted onto one plant to reduce space needs for the sake of variety—for example, apple trees which produce many kinds of apples, or single citrus trees which produce lemons, limes, and oranges. Energy-dense annuals like root crops and squashes grow easily, and though they usually tend to spread out over a wide area, can be grown vertically. Various kinds of vertical farming in general will help us to make best use of our space, especially with leafy greens and herbs. Grains like oats, buckwheat, and amaranth can be used as alternative grains.

We can also employ the use of hydroponic and aquaponic arrays which, combined with fish ponds, makes for a very effective use of space. Aquaponics is any symbiotic relationship between fish that

excrete ammonia, bacteria that convert this ammonia into nitrate, and plants that use this nitrate as fertilizer, which maximizes available resources. In comparison to traditional conventional agriculture methods, aquaponics uses only one-sixth of the water to grow up to eight times more food per acre. Because aquaponics arrays produce both a vegetable and fish crop, they provide more complete nutrition. Due to them being a closed system and the use of the fish waste as fertilizer, they also avoid the issue of chemical runoff.

Furthermore, there is a wealth of edible resources available in the ocean outside—beyond the usual fare of fish, shellfish, and molluscs, many varieties of seaweed[142] (some of which are "bacon flavored[143]") like alga[144], kelp, and other aquatic plants are delicious and nutritious. Many forms are in regular culinary use already. Kelp is also extremely promising as a renewable biofuel to produce methane and ethanol. We can farm the ocean floor extensively, which will be explored in the last section of this phase.

In short, we can grow most kinds of food, as there are relatively few crops which truly require large amounts of space. It just so happens that these crops tend to be staples of our diet, as they are the best use of wide tracts of empty land and are relatively low maintenance. The percentage of meat in our diet will have to go down, as animals take up more space and are higher maintenance. We will have to adjust our expectations of what "common" food looks like—and, with a little technology, we can streamline most of the agricultural labor.

---

[142] Wikipedia Contributors. "Edible Seaweed." Wikipedia, 6 July 2023, en.wikipedia.org/wiki/Edible_seaweed. Accessed 30 July 2023.

[143] https://www.nwnewsnetwork.org/people/tom-banse. "Whatever Happened to the Sensational Seaweed That Supposedly Tastes like Bacon?" Northwest News Network, 25 July 2019, www.nwnewsnetwork.org/food-agriculture-and-animals/2019-07-25/whatever-happened-to-the-sensational-seaweed-that-supposedly-tastes-like-bacon. Accessed 30 July 2023.

[144] Wikipedia Contributors. "Palmaria Palmata." Wikipedia, 25 June 2023, en.wikipedia.org/wiki/Palmaria_palmata#Culinary_use. Accessed 30 July 2023.

Consider placing a tax on refined sugars—they shouldn't be prohibited outright, but if sugar is more expensive, hopefully it will be used more sparingly. Honey (crystallized and milled into a solid granulated form) will make an excellent alternative. Humans are omnivores, but diet in most parts of the industrialized world usually includes a lot of meat and simple carbohydrates with a little vegetable matter and complex carbohydrates, when our bodies are designed for a lot of complex carbohydrates and vegetable matter and a little meat and simple carbohydrates. The limitations of life in an underwater city (and greater availability of vegetable rather than animal foods) will encourage this shift in balance towards a better balanced diet.

Horticulture Technology

Picture this: a Demeter gardener wants to plant a new apple tree somewhere within the city. She has a variety of tools available to her to help her in this task, including a computer loaded with a full database about plants. Perhaps she selects "apple tree" on the computer screen, like choosing a fruit at the self checkout line at a supermarket. If she is using an existing sapling imported from the surface, the computer might then ask her to put the apple tree on a scale and ask her to input the height of the sapling. From this data, the computer could direct her to a large array of dispensers to create the appropriate mixture of soil types, from layers of sterilized (to prevent fungal infection) sand, silt, clay, loam, drainage gravel, fertilizer, mulch, and compost, to be poured in with the sapling into a wheeled container—like a wagon-sized dump truck. Then, she could wheel the container to the planting site, designated by the computer to ensure neighboring plants wouldn't compete for resources. When she or her workers have to perform upkeep of the plants like pruning, checking sensors, or other work that requires them to be low to the ground, they could patrol through green spaces riding electric vehicles (like go-karts, designed and maintained by Hermes, with a zero turn radius) to reduce the strain on their backs. A sensor nodule like a tent peg could be driven into the ground near the young tree to remotely monitor soil pH, nitrogen concentration, moisture, and more. With an entire network of sensors like these, and a buried network of responsive[145] irrigation lines, Demeter will be able to monitor the conditions of the entire internal ecosystem from a single control terminal, addressing issues as they arise and making modifications as the computer recommends. They will also be able to control sunlight intensity and hours per day through a system of special LED panels. While some sunlight can be redirected from the surface through fiber optics, both human beings and plants need lots of natural sunlight to be healthy—or at least, the next best thing. Using full-spectrum LEDs, nanotechnology that mimics Rayleigh scattering, and lenticular lenses, artificial skylights convincing enough to fool a plant can compensate for cloudy days or dark corners. This also helps maintain a healthy circadian rhythm, crucial for good health. Wind should also be simulated, as a certain amount of mechanical stress is important for larger plants to grow strong roots. Perhaps signage could encourage citizens to gently shake young trees as they walk by. Any plants taller than shrubs (aside from climbing vines) should be structurally reinforced as a redundancy.

Sensors will also be able to chart planting times and give recommendations for harvest. Fruiting plants can be easily harvested (either by hand or machine) as they appear ripe, but this will be especially useful for subterranean crops like potatoes and carrots. Harvested food will be collected, washed, and prepared by Demeter in food prep facilities. Some of the food can be prepared to be eaten immediately in a public cafeteria, which will be the only source of prepared food until restaurateurs make their home in Augury. Some raw produce with a short shelf life can be made available in an arrangement similar to a farmer's market, but Demeter should maintain a circular "cannery" (not with single-use metal cans, but with reusable glass jars). Other forms of food preservation should also be used: dehydration, freeze drying, and wet/dry brining, to name a few. Sand is also a helpful preservative to store some

[145] "About - Responsive Drip Irrigation | How Plant-Responsive Irrigation Works." Responsive Drip Irrigation, www.responsivedrip.com/about-plant-responsive-irrigation/. Accessed 27 Nov. 2023.

produce in for a few months, as it keeps excess moisture away and regulates humidity. A portion of these preserves can also be provided at retail to the public. Some of the preserves, however, should be placed in circulating storage (in which a constant supply is maintained, with newer preserves being placed into storage as older preserves are taken out of circulation to be used) as an emergency food stockpile. The computer should also be able to calculate how much of each kind of crop to plant (and with a little AI, map out where they should be planted) to feed all the citizens in Augury with a reasonable amount of surplus. Any food waste or biomass scraps generated at any stage of this process can be composted and reused.

Planting beds in Augury can be designed like in-ground swimming pools; this will make it easier to control soil conditions and contain spread of disease. The bottom layer of the pool can be filled with gravel, which will serve as a water table cistern—water should be deposited directly into this layer either from dedicated supply lines or a PVC pipe running to the surface. Drainage should also be considered in case of overwatering. Atop the gravel will be placed large woody materials like logs, then branches, sticks, and mulch, then vegetable material, then compost, then topsoil to fill in all the gaps and straw/woody mulch on top to reduce dehydration. The exact composition and bacterial load of each bed should be tailored to suit its crops. Beds should be designed with plenty of room for deep rooting systems, as plants will have extended lifespans without risk of frost or burning. These distinct beds will make horticultural planning easier to compartmentalize.

With Aether's total atmospheric control we can rotate growing seasons, so that there are at least four total atriums/growing facilities with at least one in the harvest season at all times. This will stabilize harvest output year round. In addition to balancing harvest/growing cycles, this may also eliminate the issue of seasonal affective disorder. If a citizen were tired of spring, they could walk to another part of the city and be in autumn. These atmospheric conditions can also be monitored remotely by both Demeter and Aether.

Having these facilities adjacent to public spaces will give the people of Augury a greater appreciation of their food and all that goes into producing it.

Biome-Specific Atriums

With total atmospheric and environmental control, different atriums can be dedicated to specific biomes to increase yields. Many atriums will be tuned to temperate conditions—not too hot, not too cold, no extreme temperatures to burn plants. But some atriums can be attuned to rainforest or tropical conditions—higher rainfall, sunlight, temperature, and humidity—to better grow some of our favorite produce such as avocados, bananas, coconut, and melons.

Fauna

As previously stated, a complete ecosystem also includes animals. Plants benefit cohabitation with animals to pollinate, provide fertilizer, and more. Though Augury won't have room for large livestock animals like cows, pigs, or horses, smaller livestock like sheep and goats may be sustainable sources of meat, dairy, and wool. Poultry like chicken, duck, and goose, etc. are viable options, don't require too much space, provide meat, eggs, and feathers, and benefit many systems due to their diet and activities. Aquatic animals like fish, shellfish, and molluscs are extremely promising–not only are they rich in calories and nutrients, but naturally clean the water they inhabit, and can be grown vertically. Many insects are beneficial to ecosystems as pollinators and more, and are excellent sources of protein. Pollinators like bees, hummingbirds, and butterflies can be imported (though the diets of caterpillars must be carefully monitored to prevent them from becoming pests). The larger animals will require space to move around, so they can be housed in atriums. As Demeter is able to control the populations of these animals, pest problems will be minimal compared to surface gardening. Populations of carnivorous birds and bats can be maintained to control populations of pests like flies, gnats, mosquitos, etc.

Apiary

Honeybees are a precious but endangered natural resource in the world, and Augury can be an excellent refuge for them. With a centralized apiary connected via ductwork to the atriums, Demeter can maintain a local population of honeybees to pollinate plants and produce honey and wax. The aforementioned atmospheric controls will make the bee colonies less susceptible to parasites and disease. Automated mechanisms and sensors like those mentioned previously can contribute to servicing and maintaining the apiaries without disturbing the hives. Consider mechanisms similar to the patented Flow Hive to harvest excess honey noninvasively. Steps may need to be taken for close cohabitation—in time, the bees will become accustomed to staying around people, and citizens will learn not to fear them, but tourists and visitors may still pose a threat. Signage may reduce this.

Note that "honeybees are sensitive to pulsed electromagnetic fields generated by [wireless] technology," which can temporarily disorient them[146][147]. This issue is explored in Part Two: Apollo Phase 1: "Radio EMF Concerns."

Nature Preserve

Endangered and rare species of animals and plants could find a stable and protected home in Augury's atriums, where they would be safe from poaching, extreme temperature fluctuations, and habitat loss, and could be easily monitored and cared for. Some animals and plants could coexist in atriums with humans; some will need dedicated preserves to keep them separate from human interaction. Ecotourism may also be viable to help maintain these environments.

---

[146] "Cell Phone Bee Mortality Link: Sensationalism Not Science." Vanderbilt University, 14 June 2011, news.vanderbilt.edu/2011/06/14/cell-phone-bee-mortality-link-sensationalism-not-science/. Accessed 4 Aug. 2023.
[147] Jungwirth, Joseph, "The Effect of Electromagnetic Fields Produced by WiFi Routers on the Magnetite (Fe3O4) Particles of the Italian Honey Bee (Apis mellifera ssp. ligustica)" (2019). Culminating Projects in Biology. 42. https://repository.stcloudstate.edu/biol_etds/42

Consider the construction of a large "nature preserve" atrium, which will be especially beneficial in the event of partial or complete surface ecological collapse—additionally, Augury would also be an excellent facility for long-term emergency seed and embryo storage vaults.

Green Spaces

The green space atriums are crucial to the community health of Augury, not just physically, but mentally as well. On the surface, green spaces are crucial for supporting active lifestyles and improving access to exercise opportunities, fostering healthy eating habits, improving air quality, regulating temperature, reducing stress, enhancing cognitive function, encouraging positive youth development, and reducing anxiety and depression[148]; these effects will be amplified in the enclosed underwater environment.

The primary function for the atriums must be to grow beneficial plants, which either provide food or a natural resource or significantly contribute to air or water cleanliness. These plants should be arranged in companion guilds as living, balanced ecosystems rather than for appearances. However, in addition to functioning as a greenhouse, the atriums will be the common spaces of the city. There we will have our plazas, courtyards, and squares; some shops and restaurants, walking trails—places where "people gather and interact to build social cohesion and foster social capital." Landscaping should marry efficiency in planting with human pleasure and enjoyment. There should be plenty of comfortable places to sit or recline, features like gazebos, fountains, decorative ponds, and picnic tables, and fruits ready for picking. Moss and low-rise creeping plants can be used in lieu of grasses for ground cover. Green spaces must be always open and equally available to all citizens of Augury to occupy and do as they please, as long as they treat the space and

[148] Larson, Lincoln, and Aaron Hipp. "How Green Spaces Can Improve Your Health." College of Natural Resources News, 20 Apr. 2022, cnr.ncsu.edu/news/2022/04/parks-green-spaces-improve-health/. Accessed 19 Aug. 2023.

others with respect. Litterers should be punished with community service to undo the damage they caused with interest.

Ocean Agriculture

To support the food and resource needs of Augury's population, we can pursue marine agriculture by planting wide (and otherwise empty) tracts of surrounding seabed[149] with marine vegetation such as kelp, seagrass, and eelgrass (all of which pull carbon dioxide from the water, replace oxygen, and provide food and habitats for animals). In time, plants can be bred or genetically engineered to produce bigger and better crops, just as we have done historically with terrestrial crops. Some terrestrial crops may even be bred or modified to survive in saltwater. Free-floating fish farms can be maintained in netting (which must never be made of plastic, as plastic fishing nets are the deadliest form of ocean pollution[150]). "Suburban" or rural communities in smaller compounds can be constructed in the outskirts of the city for farmers to have a more direct access to agricultural zones. To prevent overfarming, overharvesting, and soil depletion, permaculture practices and biodiversity are important.   The wild animals living in these underwater farms should be hunted only by hand, not with machines or large traps. Populations should be monitored to prevent overhunting. Fortunately, permaculture will be easier to pursue, as ocean plants do not die in winter, so yearly replanting will be unnecessary. Underwater equivalents of tractors and other heavy agriculture machinery could be designed, manufactured, and maintained by Hermes.

---

[149] Marine plants can grow up to 600 feet deep (which is how deep sunlight reaches).
[150] Laguna, Alejandro. "Fishing Nets: The Double-Edged Plastic Swords in Our Ocean." UNEP - UN Environment Programme, 5 June 2023, www.unep.org/technical-highlight/fishing-nets-double-edged-plastic-swords-our-ocean. Accessed 16 Oct. 2023.

XIX. Poseidon Phase 2: Interior Water Features

Within the atriums should be freshwater features of all kinds, from taller waterfalls to small fountains, the primary function for which is simple human enjoyment. Citizens should be able to enjoy the atriums alongside babbling brooks and tranquil ponds. The water features themselves can house fish, molluscs, and freshwater plants to condition the composition of the water and provide some resources. Properly designed, these water features could be harnessed into a hydroponic array to irrigate plants. Waterfalls atomize some water vapor and contribute to humidity, which will make the local atmosphere friendlier to plants. Some of the larger water features should be kept clean enough for wading and bathing (not with soap), filtered and kept free from algae by the aforementioned plants and animals. Fountains large enough can also be used for wading. Humans love water, and being able to be near it and smell it and play in it will boost our mental health.

XX. Hephaestus Phase 2: Full Fabrication Facilities

I've mentioned several times throughout this manuscript already that everything in Augury should be designed with maintenance and longevity in mind. This is of the utmost importance. We live in an age of planned obsolescence—a repulsive business practice in which products are intentionally made at lower quality so that the consumer will buy a replacement sooner. Augury must be designed in the reverse. Everything that we make should be designed of the highest quality we are capable of—to last, to be repaired, to be used. We should embrace the Japanese principle of **Kintsugi**, which is "the general concept of highlighting or emphasizing imperfections, visualizing mends and seams as an additive or an area to celebrate

or focus on, rather than absence or missing pieces...as a philosophy, it treats breakage and repair as part of the history of an object, rather than something to disguise[151]."

Products and materials should be durable; built to last as long as possible. We must still keep expenses in mind—we don't want to overuse exotic materials which will make whatever we produce too expensive to use. Machines should be elegant, mechanically simple, easy to use, easy to make, easy to repair, and long-lasting. All of these considerations are elements of good design. Good design is what we must pursue, over profit, or we will sacrifice good design, and make someone else's life harder, in order to profit.

Technology is our power. Everything from fire to a sheet of paper to a smartphone to the electron microscope has the power to make our lives easier through modifying parts of our environment. Every age of humanity has been defined by the technology available to us—and the drunken hedonism of the industrial revolution has taken its toll on us, because we only gave thought to runaway explosive expansion, only our potential, with no consideration to the consequences. Because of this, we work hard—harder than we've ever needed to. Research indicates that the modern workers labor longer hours than medieval peasants,[152] and for what? To make big corporations bigger? To wallow in unimaginable luxury and convenience? Because we don't have any other option for our own survival than to enable CEO's and board members' and stockholders' insatiable megalomania? How many decades did it take to convince factory owners to allow workers a few days of rest, or to make child labor illegal? With the technology available to us, we can enjoy our comfort and conveniences without working all day. With the power of our technology comes responsibility—a burden we must respect. Technology should be used to make all of our lives easier, to reduce

[151] Wikipedia Contributors. "Kintsugi." Wikipedia, Wikimedia Foundation, 17 Mar. 2019, en.wikipedia.org/wiki/Kintsugi. Accessed 1 Sept. 2023.
[152] Bilyeau, Nancy. "Do You Work Longer Hours than a Medieval Peasant?" Medium, 2 Oct. 2021, tudorscribe.medium.com/do-you-work-longer-hours-than-a-medieval-peasant-17a9efe92a20. Accessed 3 June 2023.

the amount of work any of us have to do, rather than to bring wealth to a small handful of powerful individuals. With modern technology and a little cleverness, we can design an entire city which provides all of our basic needs with little need for work on our parts. We should use the technology we have—but we must always take care that our goal, our motivation in the usage of what little power we have is to help others before elevating ourselves.

Necessities should be prioritized above luxuries. It's imperative that we be capable of producing everything we need to survive, should Augury ever need to be sealed off from the outside world. That includes, above all, spare parts to keep the Executive department branches running.

Advanced Manufacturing

As Leonardo da Vinci once said, "simplicity is the ultimate sophistication." While this is true, and we should generally strive to keep our technology as simplified as possible, it's important to balance simplicity and complexity, and use the technology we have to its full potential for maximum benefit rather than hiding the best stuff away where only a few people can use them.

Conductive Ink Printing

While not suitable for any electrical wiring intended to carry significant amperage, electrically conductive ink/paint can be used for lighting or sensors, for low-voltage electronics, as EMI shielding, and be applied to almost any surface.

PCB Printing

In a technology-heavy environment like Augury, circuit boards will be in high demand—but e-waste is a growing problem. Of course, all electronics will be designed to last as long as possible, with replaceable components, but recyclability will also be important. To make this easier, Hephaestus will maintain in-house circuit board

printers, so new microchips can be easily designed and manufactured.

Prefabricated Modular Building Components

Interior structures should be designed with longevity and ease of maintenance in mind. Consider building almost everything from "plug-and-play" style prefabricated, modular, interchangeable parts, which can be mass-produced, stockpiled and easily swapped out—like the construction equivalent of Legos. Imagine a wall panel already wired for electricity with built-in outlets, removable panels for plumbing fixtures or vents, different versions to include windows or doors. If the panel was damaged or you needed to remodel, you could simply "unplug" it and "plug-in" a new one rather than struggling with running new wire or slow installation times. Controls are a good example of this. Consider how control panels in space shuttles and stations are built: buttons, levers, and dials are large, chunky, and sturdy. They are analog, durable, and easy to disassemble, repair, and replace. Touch screens are extremely convenient and versatile but fragile; if they are used, they should be designed to be interchangeable and spares (of stock sizes) should be readily available.

Some contracting companies are already starting to produce sets of standardized and prefabricated wall/floor/roof systems which are predictable in performance, cost, and delivery. The "Lego-like components" work together to create an envelope that is ready to be finished with interior and exterior materials, allowing for aesthetic and regionally appropriate treatments and customization.[153] If these panels as well as utility fixtures are composed of predictably divisible units (use metric system) as Legos are, they will fit together easily.

---

[153] "Prefab Panels." B.Public Prefab, bpublicprefab.com/prefabwholesale. Accessed 20 July 2023.

Public Workshop

Hephaestus should maintain public workshops to empower citizens of Augury to build and repair things for themselves, as well as educate industrious individuals on the tools and processes that go into manufacturing and fabrication. These workshops include workspaces specialized for woodworking, metalworking, glass blowing, electrical work, electronic design and maintenance, textiles and sewing, and more, and will host regular classes on the operation of tools and machines as well as the production process of common items.

Textiles

In our modern economy, fast fashion accounts for an enormous portion of landfills and sweatshops. Clothes are made faster, of lower quality, often out of plastic, and intended to be disposable—in my opinion, an insult to one of the oldest human technologies. Our perspective in Augury should be more sustainable. Hephaestus can maintain facilities for local manufacture of clothing, for both individuals and companies. This can include machines to spin natural fibers into string and yarn, weave fabric, and cut and sew fabrics into usable items. Some fibers like cotton and coir would have to be imported, but fibers like hemp, wool, linen, spider's silk, and chinegora (dog wool; see part 3, Mental, Emotional, & Social Health; Therapy Animals) could be produced locally. Artificial textiles like acrylic, nylon, and spandex that are made from plastics must be avoided, as they will release plastic fibers into the water with every wash that will never fully degrade and may be ingested. Some dyes and pigments can also be made locally, though toxic pigments must be avoided as they may contaminate the water supply.

Clothing

Though most people immigrating to Augury will bring their wardrobe with them, eventually people will want to buy clothes locally. To facilitate this, consider something similar to the following system. A prospective tailor applies to work in Augury, and is accepted if they can demonstrate proficiency (and if there is not a preexisting surplus). If accepted, they operate under the Hephaestus umbrella. They have the option to work from home or in Hephaestus facilities where they will have access to commercial-grade sewing machines, looms, spinning wheels and so on. The price of the garments they produce would be determined by the hours of labor and cost of materials and then sold at the public market. Larger sizes of clothing would be more expensive, but only ever enough to cover the cost of the additional material used. Hephaestus tailors could also offer services to modify and repair clothing; some may specialize. Having clothing manufacturing localized instead of outsourced and mass produced will encourage a culture of longevity and sustainability with

clothing, in contrast to the norm of replacing cheaply made, impractical clothing every season. Quality should be prioritized over quantity. This system needs further development, but I believe it shows promise.

Simpler Machines

As Leonardo da Vinci once said, "Simplicity is the ultimate sophistication." It's important not to abuse or overuse power and technology. We should avoid using complex machinery when simpler machines will do. Developing a full reliance on electricity and computerization could become an exploitable weakness, and can also weaken us physically if we completely remove all need for manual labor. This is not to say we give up on automation—our goal should still be to "work smarter, not harder." For example, clockwork mechanisms can automate mechanical devices without the need for electricity or electronics. Clever application of simple machines or manually-powered machines can utilize pulleys and levers to make operation easier. Consider also the many forms of passive design, which harness forces of nature like gravity or convection to run. Proper design of a machine on the front end will reduce its needs in practice. Like with modern homesteading, fusion of modern and primitive machinery will achieve a better balance. The simpler a machine is, the easier it can be serviced and repaired, the longer it may last, and the easier it will be to build. Similarly, avoid expensive, complex raw materials where unnecessary—Aluminum oxynitride may be the best material for exterior windows, but requires laboratory settings to work with. Glass is easier to manufacture and requires only heat to work, and should be used for interior windows.

As mentioned previously in this section, consider the efficiency of the Lego system. Not only are building blocks interchangeable, but also a complex system of gears, wheels, axels, pulleys, motors, power sources, pneumatic pistons, pins, beams, remote controls, and more—all of which are of standard sizes and compatible with each

other[154]. A similar system in Augury's machinery and parts manufacturing (with corresponding serial numbers for organization) would make repairs, upgrades, and maintenance far easier. Custom parts for each project increases expenses and makes finding spare parts more difficult.

Modular Open-Source Smart Devices

Whether we like it or not, modern culture has come to rely on smartphones. Rather than give them up entirely, I propose we improve on the concept with customization. Partnering with Minerva, Hephaestus can manufacture stock parts which can be assembled by the consumer (assembly could also be provided as a service) to make a custom smartphone. Just like with computers, the battery, touchscreen, memory, storage, and SoC (System on a Chip) could all be individually replaced or upgraded with different capacities rather than replacing the entire device. Features like cameras, ports, buttons, flashlight, microphones, antennas, input devices, modules, and sensors could be added, replaced, or removed depending on the consumer's preferences. One person may want a deluxe device loaded with all the latest gadgets, while another may prefer a simpler device which is outfitted with nothing more than a microphone, speaker, antenna, and number pad. Parents will be able to control exactly what their childrens' devices are capable of doing. Devices could be specially outfitted with things like thermal sensors or special programs to help people do their jobs. Some people will have slim, minimalistic devices while others' will be thick and powerful. This approach will greatly reduce costs and e-waste, and improve the individual's experience by giving them total control over their personal device. The operating system should be similarly modular, and perhaps even open-source. Minerva could release a standard software package, and individuals with coding skills could modify as they like. The devices should also easily interface with computers, to streamline customization. Computers, of course, should follow the

---

[154] "Lego Technic." Wikipedia, 27 Feb. 2023, en.wikipedia.org/wiki/Lego_Technic.

same principles, to maximize modularity, customization, compatibility, and reparability.

XXI. Minerva Phase 1: Automate Critical Systems

I've described at several stages in this plan so far how technology can be used to automate or streamline crucial functions of the city. That's where the Minerva department comes in. As the balanced, finished version of Augury is coming into view, Minerva will further improve the crucial functions that keep it running to make better use of our limited time and resources. As with any effective system, Minerva should have built in redundancies and failsafes. Below are a list of crucial primary operations that automated systems should be able to regularly perform without human input to keep Augury running as a habitable environment:

- Power generator monitoring
    - Basic maintenance (such as periodic application of lubricants)
    - Monitoring reactor temperature at core and moving parts
    - Revolutions of moving parts
    - Power output metrics/analytics
    - Measuring remaining fuel mass
    - Malfunction alert
    - Radiation detection and isolation
        - Emergency alarm activation
        - Emergency reactor flooding
            - With manual override
- Maintaining electrical current to exterior shell
    - Current leak detection
- Leak detection
    - Sump pump activation
    - Bulkhead activation (in case of flooding)

- o   Emergency alarm activation
- Fire detection
  - o   Detecting hot spots above a certain temperature city-wide
  - o   Deployment of extinguishing countermeasures
  - o   Emergency alarm activation
- Air quality control
  - o   Monitoring pressure, temperature, cleanliness, and composition city-wide
  - o   Pressure equalization
  - o   Carbon monoxide detection
  - o   Oxygen level detection
  - o   Algae reoxygenator maintenance
    - ■   Bubbler flow control
    - ■   Removal of waste biomass
    - ■   Fertilizer control
    - ■   Light exposure
  - o   Countermeasures to restore any imbalances
  - o   Notification of technician in event of malfunction
- Water quality control
  - o   Monitoring composition, temperature, pH city-wide
  - o   Countermeasures to restore any imbalances
  - o   Notification of technician in event of malfunction
- Ecosystem Management
  - o   Balance of growing conditions including irrigation, sunlight, soil composition, humidity, etc
  - o   Notification of technician in event of imbalance
- Healthcare
  - o   Activate emergency power (batteries, generators, etc)
- Emergency
  - o   Surface condition monitoring
  - o   Bulkhead control in case of radiation/disaster detection

All decisions and actions taken by machines should be logged in a Git-style data repository.

XXII. Minerva Phase 2: Further Automation

After the vital functions needed to keep Augury livable have been automated, we can begin to streamline our other utilities. Remember that while automation usually involves electronics and programming, it doesn't have to—the concept of automation simply means to reduce the need for human intervention. This can often be achieved with clever implementation of simple machines, which are easier to maintain and will continue to function in the event of a blackout. On the other hand, there is a lot we can achieve using artificial intelligence and machine learning, especially with jobs like those in the clerical field. In fact, AI can design code and programs to improve automation! Proper UI/UX design in equipment for each department alone will greatly increase efficiency. I have included some general ideas that have occurred to me below, each of which are well within our current technological capacity:

- General
    - Automatic door control
    - Data recording
        - To reduce time spent by humans on paperwork
    - Information
        - Information acquisition, possibly from kiosks throughout the city
        - Augmented reality
    - Hazard detection (city-wide)
        - Carbon monoxide, low/high oxygen, radiation, smoke, heat spike, water (flooding), explosive gas, etc.
        - Alarm triggering
        - Venting/circulation
    - Thermodynamics monitoring
    - Maintenance scheduling
    - Navigation
    - optimization
- Bia

- - Output monitoring
    - Analytics reporting
  - Power storage (gravity battery)
  - Light control
    - Intensity
    - Temperature
  - Ground fault circuit interruption
  - Electrical grid monitoring
    - Current flow
    - Resistance
    - Heat losses
- Poseidon
  - Water level monitoring
  - Interior water feature control
  - Filter cleaning/replacement
  - Internal leak detection
  - Microorganism monitoring
- Aether
  - HVAC conditions
  - Weather monitoring/reporting
  - Precipitation (if applicable)
  - Filter cleaning/replacement
  - Microorganism monitoring
- Hephaestus
  - Storage organization
  - Assembly line
  - Manufacturing materials/spare parts for scheduled repairs
  - Generative invention design
  - Design optimization
- Hestia
  - Structural maintenance
  - Engineering optimization
    - Structural design
- Demeter
  - Plant health monitoring
    - Diagnostic & treatment
  - Seed sowing (drones)

- Planting site selection
- Optimal layout design
- Soil composition recipe for planting beds
- Plant genetic engineering
  - Adaptations to be tolerant of lower light/higher yield
- Nutrition/ecosystem design
  - Requisition plants to be added to balance available nutrition
- Food preservation and storage
  - Store quantity reporting
  - Consumption rate monitoring
  - Phasing out old supplies
- Apollo
  - Data speed/usage monitoring
  - Research data from other departments
    - Data analysis
    - Logging
    - Analytics reporting
    - Recommendation of improvements
    - Predictive analysis
- Hermes
  - Traffic recording/analysis
  - Automated transports (drones of various sizes)
  - Supply acquisitions
  - Scheduling vehicle maintenance
  - Damage analysis
- Aegle
  - Design of prosthetics/casts
  - Trend monitoring
    - Outbreak early warning system
  - Diagnostic recommendation
  - Testing
    - Test operation
    - Results interpretation
  - Medication
    - Allocation
    - Dispensing

- Manufacture

Data from each of these processes should be logged with Apollo for human monitoring. All automations must have a force-quit manual override. **All machines should have a manual mode—no machine in Augury should ever be rendered unusable by an automation failure.** All decisions and actions taken by machines should be logged in a Git-style data repository. Some actions should require human approval before execution.

XXIII. Demeter Phase 3: Exterior Ecosystem

Humility is an important aspect in the health of a culture. Hopefully the deep, endless abyss of the ocean will help us to remember how small we are. Some consider the human race as just another animal in the ecosystem—where others see us as masters of our domain, kings of all we survey. As the saying goes, "with great power comes great responsibility[155]," or, *noblesse oblige*, and we undoubtedly have more power than any other life form on Earth. I consider the human race to be stewards; guardians of our beautiful world and all the life on it, which is of superior design quality to everything we ourselves have created or ever will create. It is better for us to cooperate with the design of nature as much as possible, rather than resist it. For the human race to relocate, in part or in whole, to the ocean, will reduce the strain that industrialization has inflicted on surface ecosystems. There is no need to displace existing ecosystems since so much of the ocean floor is empty space, but the Biorock construction will turn Augury itself into a brand new ecosystem. The Biorock mechanism cleans ocean water from dissolved minerals, adds hydrogen and oxygen to the water which encourages marine life, repels sharks due to its electric field, and is self-healing as long as the electric current remains active. Furthermore, Biorock is

---

[155] Stan Lee, "Amazing Fantasy" #15, 1962

currently used to restore coral reefs because the surface of the aggregated minerals is perfect for coral growth. "It is a unique method that allows coral reefs, and other marine ecosystems including seagrass, salt marsh, mangrove, and oyster reefs to survive and recover from damage caused by excessive nutrients, climate change, and physical destruction by greatly increasing the settlement, growth, survival, and resistance to stresses, including high temperature and pollution, of all marine organisms. As a result, it keep[s] ecosystems alive when they would otherwise die from severe stress, and restore them at record rates where there has been no natural recovery. Around 500 Biorock™ reef structures have been built in around 40 countries all around the world[156]." Therefore, the entire surface of Augury's external accreted structure will eventually grow into a coral reef, further encouraging the health of the surrounding ocean, preserving marine life by serving as an artificial marine nature preserve. Life will grow on the exterior surfaces of the city due to biofouling[157] no matter what we do; better to embrace it than fight an ongoing war against it.

"Coral reefs protect coastlines from storms and erosion, provide jobs for local communities, and offer opportunities for recreation. They are also a source of food and new medicines. Over half a billion people depend on reefs for food, income, and protection. Fishing, diving, and snorkeling on and near reefs add hundreds of millions of dollars to local businesses. The net economic value of the world's coral reefs is estimated to be nearly tens of billions of U.S. dollars per year[158]." Augury should become a beacon of appreciation, preservation and research of reef ecosystems, like an inverted aquarium. Our technology and resources can do a lot to protect coral colonies from stressors, which include pollution, acidification,

---

[156] "BiorockTM, Mineral Accretion TechnologyTM, SeamentTM." Global Coral Reef Alliance, 2009, www.globalcoral.org/biorock-coral-reef-marine-habitat-restoration/.
[157]The accumulation of microorganisms, plants, algae, or small animals on wet or submerged surfaces, especially in the ocean.
[158] NOAA. "Coral Reef Ecosystems." Www.noaa.gov, National Oceanic and Atmospheric Administration, 1 Feb. 2019, www.noaa.gov/education/resource-collections/marine-life/coral-reef-ecosystems. Accessed 26 Aug. 2023.

sedimentation, physical damage, and overfishing. With specialized facilities like a coral nursery, Augury could even become a center of coral ecosystem rehabilitation and export revitalization resources worldwide. Being surrounded by so much life will hopefully give the people of Augury a sense of perspective or ecological responsibility. Time is of the essence—"Coral reefs are in decline in the U.S. and around the world. Many scientists now believe the very existence of coral reefs may be in jeopardy unless we intensify our efforts to protect them[159]."

One of the greatest threats to ocean ecosystems is water **acidification**; as ocean water absorbs more carbon dioxide from the atmosphere, it becomes more acidic and less hospitable to marine life[160]. We can reduce this process by **planting wide tracts of seabed with marine vegetation like kelp, seagrass, and eelgrass**, all of which pull carbon dioxide from the water, replace oxygen, and provide food and habitats for animals[161].

XXIV. Hestia Phase 5: Auxiliary Space

Expansions to the city should be built in the same stages as described in this plan, staggered, so that as a section is being finished, another section is in progress. As Biorock walls take years to grow, the unfinished watertight sections of the city should be outfitted with utilities but kept empty until the next section can be sealed. These empty sections can be used for construction

---

[159] United States Environmental Protection Agency. "Threats to Coral Reefs | US EPA." US EPA, 13 Apr. 2022, www.epa.gov/coral-reefs/threats-coral-reefs. Accessed 26 Aug. 2023.
[160] NOAA. "What Is Ocean Acidification?" Noaa.gov, NOAA, 26 Feb. 2021, oceanservice.noaa.gov/facts/acidification.html. Accessed 1 Sept. 2023.
[161] Monterey Bay Aquarium. "The Ocean's Weedy Weapons against Climate Change." Montereybayaquarium.org, 2023, www.montereybayaquarium.org/stories/seagrass-kelp-help-climate-change-ocean-acidification. Accessed 1 Sept. 2023.

equipment storage, supply stockpiles, and emergency refugee lodging.

XXV. Minerva Phase 3: MDDS & Legislative Body

The Legislative body, which largely relates to accepting or denying policy change, is comprised of the entire population of Augury, facilitated by the MDDS. The underlying principle is that the operation of Augury, from the Judicial branch to the Executive, is ultimately under the control of its population. The Judicial and Executive branches exist to serve the best interest of the population and optimize the city.

Inspired by e-voting in Estonia[162], the Moderated Direct Democracy System (MDDS) is an internet-based system that makes voting as accessible as possible. The MDDS can be accessed by any citizen's personal device with an internet connection via a website or application, connecting users with their registered data with three-factor verification. This system operates in close proximity to all other internet-based municipal networking managed by Minerva and Apollo. This system removes the need for representative-based democracy and representative politicians. Citizens may undertake law-making, policy formation, and regulation enforcement.

Any proposed change in policy (typically proposed by a head of an Executive or Judicial branch) must be put to a popular vote, and all citizens should have until a specified deadline to cast their vote. Simpler policies will have shorter deadlines than greater changes, to allow voters time to explore related data to consider consequences and alternatives. If a citizen wants to propose a change in policy or

---

[162] Wikimedia Contributors. "Electronic Voting in Estonia." Wikipedia, 8 Aug. 2020, en.wikipedia.org/wiki/Electronic_voting_in_Estonia. Accessed 18 July 2023.

law, they could start a petition—with enough signatures, that proposal could be put to city-wide vote. Some decisions which are relevant to the entire city can be made unilaterally based on qualified experience—the best material to use for piping, or the best method of growing certain crops—but decisions which ultimately come down to a judgment call should be made by informed democracy of the entire affected population.

Furthermore, the MDDS could be able to enforce conditions to balance the fairness of voting conditions (which would be plainly disclaimed) including but not limited to the following examples.

The MDDS may:

- Compensate the weight of individual votes for factors such as expertise or conflict of interest.

- Not make some votes available to the entire population if the issue is only relevant to certain departments or demographics.

- Require voters to pass a comprehension quiz before voting on policy changes with a high estimated impact, as well as offer relevant background information concerning the issue.

- Offer a mandatory multiple-choice quiz after voting "no" on a policy change to explain their choice and request an (optional) suggested alternative for the intended goal (typically only on smaller changes)

The MDDS technicians operate as a sub-department of Minerva to design these vote modifiers and are constantly optimizing the system to be more effective and ensure any modifiers do not cause significant bias. The MDDS department itself is also open to constructive criticism by the general population, and all vote modifiers are clearly declared at the time of the vote and recorded for later reference. Certain conditions for designing votes will be

universally required for all proposed policy changes, such as the following:

Proposed policy changes must:

- also clearly declare the intended effect.

- declare the department that proposed them.

- be presented for vote individually. No votes may ever be cast to apply to multiple policy changes at once.

- be written in language simple enough for the average citizen to understand, and phrased as simply and concisely as possible. The MDDS department will have dedicated plain-speak auditors to ensure this.

The above conditions also apply to elected officials. While most officials are appointed by the direct superior to the position, the superior must choose at least three qualified candidates for the position and then allow citizens to vote for their preferred candidate through the MDDS. For example: if the city needed to replace the head of the Bia department, the City manager would choose at least three candidates who are willing to accept the position, have sufficient experience in electrical engineering or a related field, etc. These candidates would then 'campaign' by publicly exhibiting their qualifications in documentation not unlike a resume. Citizens would then have a certain period of time to vote for their preferred candidate, and current members of the Bia department would have a higher weighted vote compared to other citizens. The MDDS would then calculate the winner by popular vote and assign their role as head of Bia.

All votes, results, and contextual information would automatically be time-stamped and stored along with a current record of municipal Augury law by Apollo in a public change-tracking database repository similar to Git. Will this system be perfect and immune to hacking or fraud? Of course not—what system is? But I don't believe it will be

especially prone, and it will make democracy more accessible and accountable than ever.

National Media Platform

A community as technology-centered as Augury needs a centralized media platform. This platform will serve many purposes, including but not limited to:

- A portal for interacting with the MDDS
- A portal for dealing with government files such as permits, licenses, registrations, etc.
- A social media platform/forum for self expression and connecting with other citizens
- A city-wide message board for events, announcements, etc.
- A news outlet.

Privacy on this platform is paramount. Citizens must feel safe on this platform, with no concern that their data is being mined and sold. The platform should be a service provided to citizens, totally devoid of advertising except from posts made by local businesses on their own page to their followers. It should never be used to analyze the habits or preferences of citizens for any reason except for investigating a crime. That being said, the platform should also not be anonymous—users must feel accountable for their actions online. The platform should also employ no optimization algorithms towards content prioritization, and should actively avoid the tactics used to make social media addictive.[163]

---

[163] Miller, Sarah. "The Addictiveness of Social Media: How Teens Get Hooked | Jefferson Health." Www.jeffersonhealth.org, 2 June 2022, www.jeffersonhealth.org/your-health/living-well/the-addictiveness-of-social-media-how-teens-get-hooked. Accessed 18 July 2023.

Digital Fiat Currency

One of the requirements for a state's sovereignty is possession of an independent currency. Augury's economy should use a centralized digital fiat currency[164] called **Aurichalcum**[165] (referred to as copper or coppers for short), the blockchain data for which will be stored on Apollo servers, but moderated and monitored by Minerva algorithms with as little human interaction as possible. Minerva systems will keep an overview of Aurichalcum units and their ownership and define whether new units can be created. If new units can be created, Minerva defines the circumstances of their origin and how to determine the ownership of these new units. If Minerva detects any issues, significant value fluctuations, or problematic trends, it will alert a human technician. Though Minerva tracks the movement of Aurichalcum movements, it would be illegal for a human technician to track exchanges of currency between individuals without a Judicial warrant. Human economists should monitor this system to ensure economic health as a redundancy, especially for the first decade or so until the Minerva algorithms demonstrate effectiveness.

Citizens should be issued an ID card with a secure RFID chip which can be used as identification, a debit/credit card, passkey for certain doors, and redemption for online purchases or tickets, and certain perks afforded to citizens (for example, consider certain passageways unlockable only to citizens so they can avoid large crowds and heavy traffic from tourism). Uses of this card can be tracked only with a court order to protect privacy. The same credentials could also pair with a QR code loaded on their phone from their online government account. Of course, a fraud agency in Minerva to address identity theft would be necessary.

---

[164] Rodeck, David. "Digital Currency: The Future of Your Money." Forbes Advisor, 31 Mar. 2021, www.forbes.com/advisor/investing/cryptocurrency/digital-currency/. Accessed 18 July 2023.
[165] "The Internet Classics Archive | Critias by Plato". classics.mit.edu. Paragraph 13. Retrieved 17 November 2021.

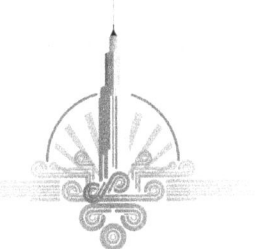

Part Three: Culture of Health

Augury should not be described as a paradise or utopia; the concept does not fit the definition of either word. A utopia, by definition, is imaginary—and is innately characterized by perfection. The primary definition of "paradise" has a similar connotation of perfection, as the concept of heaven. Make no mistake: Augury will never be perfect because people cannot be perfect. If our expectation is perfection, we will be constantly disappointed and disillusioned at best, and at worst, we will create a dystopia in a misguided attempt to achieve perfection through means of control.

The secondary definition of "paradise," however, suits Augury quite well: "an ideal or idyllic place." Ideal (or optimized) conditions are a worthy goal to reach for, though never a prize we will hold in our hand. It will take work—ongoing, never ending work in the pursuit of the ideal habitative environment for humans. The core of this environment is our culture, what it values, what it prioritizes, what it tolerates, and what our norms are.

So—how do you create culture?

The writer Will Durant interpreted Aristotle's teachings as such: "We are what we repeatedly do.[166]" Consider this in tandem with a quote by Coach Tony Blauer: "Be careful what you practice, you might get really good at the wrong thing." The only way for us to have a culture of health is for every one of us to repeatedly agree upon it. We must work together, diligently and daily, to pursue our own health and the health of others, facing problems where they develop head-on and without hesitancy. This is important at every level of society—if we do not prioritize the health of the weakest and most vulnerable of us, but show preference to the strongest and most powerful, we become hypocrites. We must consistently advocate for each and every member of our community, especially those who are unable to advocate for themselves. There is no one-size-fits-all answer to personal health. Some need food and housing. Some need therapy or healthcare. Some simply need a sense of purpose, or the opportunity to serve others. Some need nothing more than time and rest. How can we provide the same things for all people, when any amount of support will be too much (and therefore wasteful) for some and not enough for others? We must strive for equity in addition to equality. We must actively pursue success for each of us; we cannot neglect anyone. We must not be selfishly satisfied thriving while our neighbors are struggling. We can tolerate having no outcasts. No rejects. No ghettos, slums, or poor neighborhoods. We cannot tolerate intolerance and discrimination, for we are all as one. A problem in any part of our community is the responsibility of us all. When the health of all of our people is high, the health of our community will be high. We must choose to pull together—repeatedly. Constantly.

So how do we care for others? Think about someone close to your heart. How do you care for them? Do you make sure they have enough to eat? Do you care if they have somewhere to sleep, and would you take them in? If they were sick, would you care for them? If they were in trouble, would you visit them? Are you a listening ear

---

[166] Durant, Will. The Story of Philosophy : The Lives and Opinions of the Great Philosophers of the Western World. 1926. New York, N.Y., Simon And Schuster, p. 87.

or a shoulder to cry on for them? If they are incapable or ignorant, do you help them learn and grow? Do you include them in your social life? Do you give them help when they need it? Do you sacrifice for them? So it should be for all our fellow human beings. Not everyone to the same extent, of course, as we hold special fondness for family and friends—but we must not be strangers to our neighbors. This is the fundamental purpose of Augury: altruism. There is a real (though unmeasured) threshold of true care and compassion we are able to extend to those around us before triggering a sense of defensive tribalism to break down our communities into more digestible portions; for this reason, Augury should remain a relatively small community,[167] never more than maybe five thousand in population for a single colony. We have to be aware of our capabilities and limits, so that we can set attainable goals.

There are consequences to not caring for yourself. If you consistently neglect your physical, mental, and spiritual health, if you push and push and don't give yourself time to rest and recover—you are borrowing from the future, and eventually the future won't have anything left to borrow from. Burnout can affect any facet of your health and hit you without warning, incapacitating you partially or fully, for anywhere from days to months to years. If you don't schedule regular system maintenance, for the most important system in your life, eventually your system will schedule it for you.

---

[167] Savage, Tim. "The Benefits of Living in a Small Town." Parrys Estate Agents, 12 Nov. 2018, parrys.com/the-benefits-of-living-in-a-small-town/. Accessed 21 May 2023.

Our Fundamental Needs

In 1943, Abraham Maslow proposed a hierarchy of human needs[168], arranged in a pyramid with survival needs at the base and self-actualization at the top. His theory posited that human motivation is not just concerned with tension reduction and survival—but also with growth and development. He suggested that once we take care of our lower, more basic needs, we are more willing and more able to pursue higher needs. I believe the same. We must care for the basic needs, not just of ourselves, but of each other, in order to be a healthier, kinder, more moral culture. Combine this hierarchy with the Japanese concept of "**Ikigai,**" where our passion, aptitude, livelihood, and sense of purpose overlap.

---

[168] Mcleod, Saul. "Maslow's Hierarchy of Needs." Simply Psychology, Simply Psychology, 21 Mar. 2023, simplypsychology.org/maslow.html. Accessed 5 Apr. 2023.

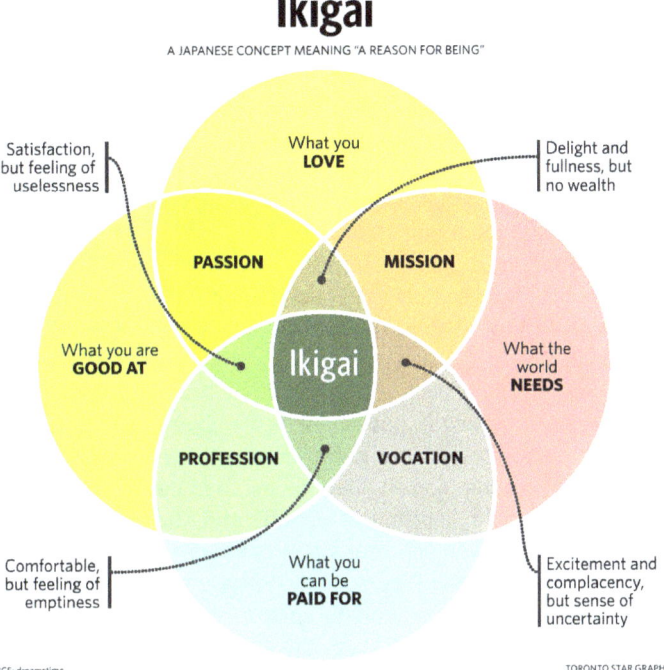

Ikigai
A JAPANESE CONCEPT MEANING "A REASON FOR BEING"

Satisfaction, but feeling of uselessness

What you LOVE

Delight and fullness, but no wealth

PASSION

MISSION

What you are GOOD AT

Ikigai

What the world NEEDS

PROFESSION

VOCATION

Comfortable, but feeling of emptiness

What you can be PAID FOR

Excitement and complacency, but sense of uncertainty

SOURCE: dreamstime

TORONTO STAR GRAPHIC

If any of these four aspects fall out of balance, we ourselves feel unbalanced. But when these needs are fulfilled, we feel fulfilled, and complete. At the core of our beings, we have many needs, all of which must be balanced with the rest. But are we capable of meeting all of these needs adequately on our own? We must tend to all of them, but in a society of people, the full responsibility of needs fulfillment cannot fall on the individual alone. We need a shift in fundamental perspective and priority, and revisit what the social structure we live in is for.

What should the structure of a society do for us? We choose to gather together in mass settlements and colonies rather than fend for

ourselves in the wilderness—how must a society serve its constituents to justify itself? What *is* a society, technically speaking? Most simply: "a voluntary association of individuals for common ends[169]." I identify two crucial parts to this definition: "voluntary" and "common ends." Voluntary—meaning we choose it. Why would we willingly persist in an association that fails to benefit us? If we choose to make the association—why would we not make it in such a way that collectively benefits us all as much as possible? If our mode of interaction merely centers around competition, why cohabitate at all? We live close to each other to benefit from each other, to operate as a team. To pursue common ends. The word "community" has a similar definition as society when used in one sense, but in another, it is "a feeling of fellowship with others, as a result of sharing common attitudes, interests, and goals."

In an excellent article[170] published in the Open Journal of Social Sciences, authors Hossain and Ali describe the relationship between an individual and the society they inhabit, and describe the ultimate goal of society to "**promote good and happy life for its individuals**. It creates conditions and opportunities for the all round development of individual personality. Society ensures harmony and cooperation among individuals in spite of their occasional conflicts and tensions." They go on to assert that "a human being…cannot live without association. So man's life is to an enormous extent a group life. Because individuals cannot be understood apart from their relations with one another," and " [a human] has to live in society for his existence and welfare. In almost all aspects of his life he feels the need of society. Biologically and psychologically he is compelled to live in society." An ugly picture begins to emerge when we consider this and see how isolated we are now, the loneliness we suffer from. The community we need as human beings is slowly killing us. As long as we are in our current culture, health will never be a

[169] Merriam-Webster. "Definition of 'society.'" Merriam-Webster.com, 2019, www.merriam-webster.com/dictionary/society. Accessed 11 Apr. 2023.
[170] Hossain, F. M. Anayet, and Md. Korban Ali. "Relation between Individual and Society." Open Journal of Social Sciences, vol. 02, no. 08, 14 Aug. 2014, pp. 130–137, file.scirp.org/Html/8-1760197_49227.htm, https://doi.org/10.4236/jss.2014.28019. Accessed 11 Apr. 2023.

priority—the profits of industries such as pharmaceuticals, energy drinks, entertainment, pornography, tobacco, alcohol, and more will never allow it.

"What We Owe to Each Other[171]" is the focus of Professor T.M. Scanlon's (excellent) and best-known book, and a crucial question we must ask ourselves if we choose to live among other people. Some believe in individualism or libertarianism and assume no responsibility for the wellbeing of others. But I believe in the obligation of the social contract[172]. I believe to be a healthy community we must each think beyond ourselves, not be narrow-minded in selfishness, but look to the wellbeing of others and the community at large as a priority. The collaborative nature of community requires us to adopt a delicate balance in our mindset—we are obligated to contribute what we can, but without a transactional expectation of others to do the same. Instead, we should do the best we can—serve the common good and others, with or without thanks or reciprocation. In kind, we must care for our neighbors whether or not they are able to compensate us or support themselves unassisted. As R. Buckminster Fuller wrote to New York magazine in the late 1900s,

> *"We must do away with the absolutely specious notion that everybody has to earn a living. It is a fact today that one in ten thousand of us can make a technological breakthrough capable of supporting all the rest. We keep inventing jobs because of this false idea that everybody has to be employed...because, according to Malthusian-Darwinian theory, he must justify his right to exist...The true business of people should be to go back to school and think about whatever it was they were thinking about before somebody came along and told them they had to earn a living."*

---

[171] Scanlon, Thomas M. What We Owe to Each Other. Cambridge, Mass., Belknap Press Of Harvard University Press, 1998.
[172] McCombs School of Business. "Social Contract Theory." Ethics Unwrapped, 2022, ethicsunwrapped.utexas.edu/glossary/social-contract-theory. Accessed 11 Apr. 2023.

Our technology has advanced exponentially since Fuller's time, and this goal has become all the more attainable. The abilities of the strongest and most powerful are sufficient to support the weakest and most feeble—and we, like the ancient Athenians[173], should consider the opportunity an honor. Will some lazy people take advantage of such a system? Of course they will. Someone will always take advantage of any system, but that is no reason to punish those who are struggling. With all our achievements, can we boast for anything while our poor, addicts, disabled, elderly, and orphaned still struggle? Are they not still people? Are you worth more than they are because you are not struggling? "Let the person who has never done anything wrong cast the first stone.[174]"

The truth is: the vast majority of us don't have to struggle to survive anymore. We have advanced to the point that we can produce excess of our needs with progressively less input—which enables us to pursue higher goals.

For this reason, the culture of Augury will be audited and safeguarded by the three Judicial departments, to fulfill the hierarchical needs of all people—physical needs; mental, social, and emotional needs; and spiritual needs. What we treat as a priority will define our culture. How we treat each other will define our community.

"Love isn't a state of perfect caring. It is an active noun like 'struggle.' To love someone is to strive to accept that person exactly the way he or she is, right here and now."

—*Mister Fred Rogers*

---

[173] Martin, Thomas. "Only the Richest Ancient Athenians Paid Taxes – and They Bragged about It." The Conversation, 3 Nov. 2020, theconversation.com/only-the-richest-ancient-athenians-paid-taxes-and-they-bragged-about-it-147249. Accessed 11 Apr. 2023.
[174] Bar Zebedee, John. Holy Bible: Containing the Old and New Testaments: King James Version. 100 AD. Vol. 43, New York, American Bible Society, Gospel of John 8:7.

Public Service Announcements

Much of the design of Augury's culture will be unfamiliar to immigrants. To encourage new cultural norms, consider a campaign of simple animated PSA's to demonstrate appropriate public behavior and etiquette and manage expectations. Manipulative imagery or phrasing should be carefully avoided, as this could easily turn into propaganda. Data-driven sources should be cited where appropriate.

When new residents emigrate to Augury, they should be required to undergo induction training which will familiarize them with cultural norms, as well as Sign Language and emergency protocols.

Simpler Living

No doubt some readers will find the inclusion of this section strange. In an underwater facility with submerged data centers and self-healing walls powered by nuclear energy, one would expect life to be complex, filled with technical data and specialized knowledge. But in fact, proper application of technology is exactly what will enable us to live more simply. I refer primarily to bureaucracy. American culture is tainted by the influences of bureaucracy, at the expense of making life in a community far more difficult than it needs to be, mostly due to the complexity of our law and justice systems. I know people who will not let their childrens' friends play in their yard, because if a child gets hurt they as the homeowner are held liable by insurance companies. The concept of "legal liability," a corruption of accountability, prevents us from normal interactions with others. People won't say "sorry" when they are at fault because it assumes liability in court. Codes and zoning prevent efficient civil design. We cannot become truly self-sufficient: we cannot grow our own food, wire our own network service, build our own houses, run our own businesses, or harvest electricity from wind and rivers without government fees and interference. You can hardly do anything in America without having to fill out a form for the government (except buy a gun). And even if we could become completely self-sufficient, we do not truly own our own land: you still have to pay property taxes, and if you don't, your land and property are taken back.

Government organizations themselves are so encumbered at every level by documentation and oversight that they are rendered ineffective. Laws and regulations have become nothing more than obnoxious hoops to jump through, boxes to check, without exerting much real power. Lawyers make a living taking advantage of people who don't know the law as well as they do (because most people have better things to do than decipher our disaster of a legal code) and sue them for all they're worth. Then, on the other hand, gigantic corporations like McDonalds get away with unsafe business practices and turn their victims into laughingstock with corporate propaganda[175].

Our rules are not strong enough to stop the powerful, but for the rest of us, they are an inscrutable mess that can bankrupt or imprison you if you get on the wrong side of it. I do not know what the perfect or optimal legal system is. But I do know that a code so massive that the average person has no hope of ever understanding it is not the answer. I believe the key is simplicity—laws have their place, but there are always loopholes. The structure of our legal system has reduced justice not to what is right or wrong, but who has the better lawyers. Regardless, the strength in our community will not be the exhaustiveness of our laws, but in the relationships of our community. We must share a sense of honor: to apologize when we make a mistake, and to right our wrongs. This is another reason to keep Augury small: we should not be anonymous in our community. We need to know the people we live and work with, so that we are less willing to take advantage of them, and more likely to show them compassion. We need to know the people at every level of our organization hierarchy, so no one in power can hide behind a title or a corporation. We need real accountability, and it needs to be personal.

---

[175] Burtka, Allison. "The American Museum of Tort Law." The American Museum of Tort Law, 21 Sept. 2018, www.tortmuseum.org/liebeck-v-mcdonalds/. Accessed 26 Aug. 2023.

A. Physical Health

As previously stated, it's vitally crucial that we establish a hierarchy of health. Without a foundation of strong physical health, it will be harder to fight factors that threaten our mental health. We cannot adequately treat anxiety and depression with medication if the cause of that anxiety and depression is caused by lack of adequate nutrition or housing.

Before exploring the ways we can encourage a community of good physical health, let's explore the legitimacy of the issue by establishing a clear mental image of poor physical health. The easiest indication in our society of poor physical health, I believe, is the pharmaceutical industry. Our bodies are poorly cared for, and struggle to maintain themselves without additional drugs, supplements, and so on. One of the most prevalent examples of this is the popularity of energy drinks and coffee. Poor diet, low activity level, and chronic sleep deprivation leads to low energy levels; as such, there's a huge demand for beverages loaded with sugar and caffeine to keep our energy up—though most people don't even consider caffeine a drug. This parallels countless other products—fad diets and exercise routines, diet pills, muscle stimulators, and cleansing—all designed to compensate for an unhealthy lifestyle without actually doing the real work. These options can seem appealing at first, but almost always take a deeper toll later in life. Consider the pervasive health conditions caused by an unhealthy lifestyle: obesity, lung cancer, cardiovascular diseases (CVD), stroke, diabetes and certain forms of cancer can be considered "lifestyle diseases[176]." An unhealthy lifestyle weakens the body over time, making one more susceptible to disease and other conditions.

---

[176] Tabish, S A. "Lifestyle Diseases: Consequences, Characteristics, Causes and Control." Journal of Cardiology & Current Research, vol. 9, no. 3, 21 July 2017, medcraveonline.com/JCCR/JCCR-09-00326, https://doi.org/10.15406/jccr.2017.09.00326. Accessed 30 July 2023.

Healthcare

Healthcare in America is an inscrutable tennis match of subterfuge and manipulation between for-profit healthcare providers and health insurance companies, where providers are trying to collect as much money as possible from insurance companies who endeavor to pay as little money as possible, to the detriment of the patient who must pay them both. Nearly 45,000[177] annual preventable deaths are associated with lack of healthcare coverage, even patients who survive (assuming their insurance even approves the treatment they need) may be financially decimated after treatment. Around 1 in 10 adults suffer from significant medical debt, with a total of at least $195 billion across the nation[178]. This system benefits insurance companies, who report earnings in the tens of billions[179], and healthcare and pharmacy executives—not people. Why do we tolerate a system that requires us, as adults, to obtain permission for medical treatment from an entity that only exists to use the money we pay them to pay for our medical treatment?

Public health and medical care systems have been successfully implemented in most industrialized nations worldwide and should be implemented in Augury as well. Healthcare should be seen as service to help people, not an opportunity for profit—healthcare professionals who believe otherwise are not suited for healthcare. Furthermore, illness, injury, chronic disorders, and medical debt can quickly demoralize a population. Anxiety and depression about not receiving sufficient healthcare for any reason increases stress, and

[177] Cecere, David. "New Study Finds 45,000 Deaths Annually Linked to Lack of Health Coverage." Harvard Gazette, Harvard Gazette, 17 Sept. 2009, news.harvard.edu/gazette/story/2009/09/new-study-finds-45000-deaths-annually-linked-to-lack-of-health-coverage/. Accessed 27 July 2023.
[178] Rae, Matthew, et al. "The Burden of Medical Debt in the United States." Peterson-KFF Health System Tracker, 10 Mar. 2022, www.healthsystemtracker.org/brief/the-burden-of-medical-debt-in-the-united-states/. Accessed 27 July 2023.
[179] Emerson, Jakob. ""The House Always Wins": Insurers' Record Profits Clash with Hospitals' Hardship." Www.beckerspayer.com, 3 Jan. 2023, www.beckerspayer.com/payer/the-house-always-wins-health-systems-face-worst-finances-in-decades-as-payers-rake-in-record-profits.html. Accessed 27 July 2023.

stress reduces health—increasing the cost of maintaining health city-wide. Wealth should not entitle a person to superior health over another. Healthcare and medicine in Augury should be a collaborative guarantee to all citizens, just like clean water, clean air, electricity, food, and shelter.

Stress Management

Stress is the most deadly and most underestimated force influencing our health today. Stress is defined as anything "that disturbs homeostasis. It [can] be physical, psychological or social...There is no doubt that some stress is necessary to compel us to move and perform...But, [chronic] stress can take a toll on the cardiovascular, immune, memory & reproductive systems and age us faster."[180] We are facing an epidemic of untreated chronic stress today in America. It makes us weaker, more tired, less patient, more sickly, more anxious and depressed, and less willing to improve ourselves. For our higher priorities, we sacrifice the most basic things that determine our health: diet, hydration, sleep, and exercise. We constantly compromise, always for a more immediate solution to our needs rather than a long-term investment. To reduce our stress and improve our health, we need to start with a strong foundation.

Diet

"Good nutrition is essential to keeping current and future generations healthy across the lifespan."[181] We have categorically bad eating habits in America, largely due to the commercialization and industrialization of the fast food industry. Food is fattier, sugarier,

---

[180] Azab, Ph.D., Marwa. "Can Stress Kill You? What Doesn't Kill You, Kills You Slowly." Psychology Today, 26 Jan. 2019, www.psychologytoday.com/us/blog/neuroscience-in-everyday-life/201901/can-stress-kill-you-what-doesnt-kill-you-kills-you-slowly. Accessed 1 May 2023.
[181] Centers for Disease Control and Prevention. "Poor Nutrition." Centers for Disease Control and Prevention, 8 Sept. 2022, www.cdc.gov/chronicdisease/resources/publications/factsheets/nutrition.htm. Accessed 1 May 2023.

saltier, faster, and fattier than ever.[182] Food is designed to work like a drug, hitting the pleasure sensors in your brain hard and fast, making you crave more and more. A single meal from most restaurants contains more than enough calories for an entire day. Poor nutrition leads to obesity, heart disease and stroke, diabetes, cancer, all of which put enormous stress on the body. But in addition to the big, scary conditions, poor nutrition causes a lot of smaller issues which make your everyday life more difficult. Take sugar—addictive properties similar to cocaine, withdrawal and relapse symptoms included,[183] and the average American consumes over 3 times the recommended amount per day.[184] The effects of chronic overindulgence are prevalent in our society: worsened chronic anxiety and depression, fogginess, decrease in intelligence,an afternoon crash or slump, poor dental health, joint pain, poor skin health (good for the dermatology industry), greater stress on the liver, heart, pancreas, kidneys, and of course, weight.[185] Sugars (simple carbohydrates) are almost pure chemical energy, far more than your body can use—so it has nowhere to go but be stored as unhealthy fat. But how are we supposed to avoid it? Sugar is in almost everything these days. Even products that claim to be "sugar free" are not in fact sugar free. The product is merely devoid of cane sugar (sucrose), which is replaced with aspartame, sucralose, fructose, glucose, glycosides, and acesulfame potassium—all of

---

[182] Haseltine, William A. "Salt, Fat and Sugar: How Americans Became Addicted to Eating." Forbes, 29 Oct. 2022, www.forbes.com/sites/williamhaseltine/2022/10/29/salt-fat-and-sugar-how-americans-became-addicted-to-eating/?sh=6c86f6bd6b8f. Accessed 1 May 2023.
[183] Detrano, Joseph. "Sugar Addiction: More Serious than You Think | Center of Alcohol & Substance Use Studies." Alcoholstudies.rutgers.edu, alcoholstudies.rutgers.edu/sugar-addiction-more-serious-than-you-think/. Accessed 1 May 2023.
[184] "Added Sugar." The Nutrition Source, Harvard T.H. Chain School of Public Health, 5 Aug. 2013, www.hsph.harvard.edu/nutritionsource/carbohydrates/added-sugar-in-the-diet/. Accessed 1 May 2023.
[185] Hughes, Locke. "How Does Too Much Sugar Affect Your Body?" WebMD, 6 Apr. 2022, www.webmd.com/diabetes/features/how-sugar-affects-your-body. Accessed 1 May 2023.

which are still sugars, simple carbohydrates, and are still habit-forming and high in chemical energy (as a basic rule of thumb, if something tastes sweet, it has a sugar in it). After all, why wouldn't a company make a product addictive if their only goal was to sell more of it?[186]

There are lots of substances that aren't good for us added to our food in order to boost profits in the food industry—whether to improve palatability, shelf life, or as a "safer" alternative to a regulated ingredient.[187] We fill our stomach with tasty cheap stuff and leave out the vitamins, minerals, and nutrients our bodies need. The short version is this: our food is overprocessed. I think most of us are aware of this. The issues with industrialized international food are obvious:food rots when it is being eaten by something else, namely bacteria and fungi. In order to prevent it from rotting, we have to make it less edible. Sometimes we achieve this by keeping it cold or hot or removing the water, or soaking it in something inedible to other organisms (like salt or vinegar). But sometimes, we remove some of the ingredients that make it edible and nutritious. Take flour for example—whole wheat flour spoils much faster than white wheat flour because it has more ingredients rich in fiber, nutrients and oils that spoil faster and are more sensitive to light, moisture and air. To make bread and flour that is more tolerable to manufacture, shipping, and shelf storage, it is made less nutritious. This benefits profit margins in exchange for bread that is not much more nutritionally valuable than candy.

We're better off eating food that has been minimally modified from its original version. We need fresh, raw food, harvested locally and preserved simply. Will it be as easy to work with? Will it be as tasty? No, it will not. But if we want to improve our health, we have to stop

[186] Claeson, Hanna. "The Untold Truth of Sugar." Mashed, 11 July 2020, www.mashed.com/225569/the-untold-truth-of-sugar/. Accessed 1 May 2023.
[187] Wellness Team. "What's Worse for You: Sugar or Artificial Sweetener?" Health Essentials from Cleveland Clinic, Health Essentials from Cleveland Clinic, 15 Jan. 2018, health.clevelandclinic.org/whats-worse-sugar-or-artificial-sweetener/. Accessed 1 May 2023.

taking the easy way out. We have to stop prioritizing immediate rewards, and start pursuing long-term investment. This will only be possible through community cooperation; fresh, unprocessed food is expensive and can quickly become a wealth barrier.

In Augury, the Demeter department will partner with Physicality to provide the population with a nutritious, unrefined, and varied diet. Plants are underutilized; we typically eat a tiny percentage of the edible plant species on earth—only around 0.2% of over 50,000 known. Demeter will ensure that we improve our local biodiversity. Take dandelion, for example: usually considered a mere weed, the plant is easy to grow, and propagate, edible, and is high in antioxidants, reduces cholesterol, is antiinflammatory, regulates blood pressure, and more[188]. Not only will these plants be harvested for our consumption, but these plants will be growing in public spaces, ready to be freely harvested by any citizens. Humans are omnivores, but diet in most parts of the industrialized world usually includes a lot of meat and simple carbohydrates with only a little vegetable matter and complex carbohydrates, when our bodies are designed for a lot of complex carbohydrates and vegetable matter and a little meat and simple carbohydrates. The limitations of life in an underwater city will hopefully encourage this shift in balance towards a better balanced diet. We can also encourage more meals per day with smaller portion sizes for each meal, which is better for our digestion and overall health[189].

Obesity Disclaimer

Obesity is a worldwide health epidemic that began in the USA around 1976-1980, largely caused by ultra-processed foods (UPFs) and sugar-sweetened beverages (SSBs) like soda and energy

---

[188] Fletcher, Jenna. "Dandelion: Health Benefits, Research, and Side Effects." Www.medicalnewstoday.com, 9 Oct. 2023, www.medicalnewstoday.com/articles/324083#10-possible-health-benefits. Accessed 28 Oct. 2023.
[189] DeSoto, Lindsey. "Meal Frequency and Portion Size: What to Know." Www.medicalnewstoday.com, 17 July 2022, www.medicalnewstoday.com/articles/is-it-better-to-eat-several-small-meals-or -fewer-larger-ones. Accessed 23 Oct. 2023.

drinks[190]. More sedentary lifestyles due to advanced industrialization has also contributed to widespread physical degeneration. Unfortunately, obesity has become a polarizing topic, as many people (especially medical professionals) have misapplied preconceived notions shaded by the concept of the BMI to make generalizations about anyone with a certain amount of body fat, regardless of the cause. In the words of a dear friend of mine misdiagnosed with obesity despite an extremely health lifestyle,

> *"Obesity is a health epidemic that has unfortunately been weaponized and misapplied [in the context of] public health. Instead of being recognized as a symptom of a larger and deeper issue, it is often treated shallowly, resulting in misdiagnosed patients and untreated health problems. It is a well known fact that weight issues can be linked to poor nourishment or [an] inactive lifestyle. It is easy to blame the patient for poor choices rather than investigate if they have access to healthy food and are able to exercise regularly or if they have an underlying health problem that makes weight loss a long term issue. Obesity is a multifaceted health concern that should be treated individually rather than with a one-size-fits-all perspective[191]."*

In short, obesity is a **symptom** of a variety of health issues, some of which are controllable, and some of which are not. We will be far more successful in improving public health if we address the deeper fundamental issues of poor health like inactivity, malnourishment and gluttony rather than only looking at the surface. Besides—humans come in all shapes and sizes. Body fat is just stored energy—how much accumulates is largely genetic, and not an indicator of poor health (or poor virtue) by itself. In fact, data[192] suggests that having a

---

[190] Temple, Norman J. "The Origins of the Obesity Epidemic in the USA–Lessons for Today." Nutrients, vol. 14, no. 20, 12 Oct. 2022, p. 4253, https://doi.org/10.3390/nu14204253.
[191] Ravven White, 2023.
[192] Brown, Harriet. "Obesity Paradox: Scientists Now Think That Being Overweight Can Protect Your Health." Quartz, 17 Nov. 2015,

171

higher fat content in your body is not always a bad thing—fat serves as a cushion, better absorbing the stress from injury, illness, and environmental stressors. There's enough information to write several books on the subject, but in short, there's good fat and bad fat—bad fat comes from inactivity, overeating, and poor diet. Good fat is insulation, energy storage, and a tool your body needs to survive and function properly.

Hydration

Many people think that a cup of coffee in the morning, a soda in the afternoon, and a glass of wine or beer at night is enough water for a day. Those things all have water in them, right? And if I don't feel thirsty, I must be fine!

Acute dehydration is what most people think of—a severe condition characterized by a dry mouth and throat, fatigue, dizziness, confusion, and perhaps crawling through a desert. But what most Americans suffer from is a mild chronic dehydration—still a significant stressor on the body. "Water is your body's principal chemical component and makes up about 50% to 70% of your body weight. Your body depends on water to survive. Every cell, tissue and organ in your body needs water to work properly—" and it uses a few liters per day (depending on your size and activity level)[193]. If you don't drink as much water as your body uses in a day, you're dehydrated. When was the last time you drank a few liters of water in a single day?

If you're not getting enough water, your body has trouble maintaining your body temperature, cushioning your joints, removing waste, maintaining your skin health, keeping your energy levels up, and protecting sensitive tissues. All of these conditions stress your body

---

qz.com/550527/obesity-paradox-scientists-now-think-that-being-overweight-is -sometimes-good-for-your-health. Accessed 4 Sept. 2023.
[193] Mayo Clinic. "Water: How Much Should You Drink Every Day?" Mayo Clinic, 14 Oct. 2020, www.mayoclinic.org/healthy-lifestyle/nutrition-and-healthy-eating/in-depth/wat er/art-20044256 . Accessed 1 May 2023.

and cause other systems to work harder to compensate. So, why don't we drink enough water? The answer I've heard overwhelmingly (and gave myself, as a child) is: it doesn't taste good. We're so used to everything having a strong, pleasant taste that we take free, clean water for granted. With time, we can shift culture away from this sense of entitlement and learn to value water for the precious resource it is.

In Augury, we can encourage better hydration in many ways. The sounds of water features in green spaces may trigger thirst reflexes. Since citizens will have to walk as their main form of transportation around the city, their thirst reflex with be bolstered. Filtered water fountains with bottle filling stations should be readily available, and reusable glass and metal bottles of many different kinds and styles should be manufactured by Hephaestus, and encouraged as a standard accessory for all citizens. Disposable water bottles should not be permitted.

Exercise

Our bodies are like cars in that they need to be in motion and degrade faster if left idle for long periods. Modern life lends itself to a mostly sedentary lifestyle, sitting at desks and screens all day, with physical exercise as an activity we have to go out of our way to pursue. "Being physically active can improve your brain health, help manage weight, reduce the risk of disease, strengthen bones and muscles, and improve your ability to do everyday activities. Adults who sit less and do any amount of moderate-to-vigorous physical activity gain some health benefits. Only a few lifestyle choices have as large an impact on your chronic health as physical activity."[194] Our modern lifestyle in industrialized nations lends itself to a sedentary lifestyle, so much so that office culture is turning to standing desks, office yoga, and sitting on exercise balls instead of chairs. The culture and construction itself of Augury must encourage higher

---

[194] Centers for Disease Control and Prevention. "Benefits of Physical Activity." Centers for Disease Control and Prevention, CDC, 16 June 2022, www.cdc.gov/physicalactivity/basics/pa-health/. Accessed 7 July 2023.

physical activity. As mentioned previously, Augury will be designed with minimal powered vehicles; walking and using bicycles for travel will be a big step in the right direction. Without car dependency, Augury will suit the needs of people rather than the needs of automobile infrastructure, and will therefore have the benefits of any walkable city (which also includes greater cultural engagement).

Social Interactions on Three Streets - Neighboring and Visiting

195

These benefits include reduced risk of obesity, premature death, high cholesterol, chronic illnesses, heart disease, improved blood circulation, higher self-confidence, increased bone density and joint health, better coordination, decreased depression and anxiety and better sleep. Conditions like fear for personal safety, unfamiliarity with routes, traffic noise and fumes, poor path maintenance, unpleasant aesthetics en route, and lack of lighting tend to cause pedestrian aversion and should be avoided/ compensated for[196].

---

[195] "How Traffic Alters the Social Life of Streets," 1972 study, Donald Appleyard.
[196] Dickinson, Karl. "Walk This Way! Walking Facts & Figures | CityChangers.org." CityChangers.org – Home Base for Urban Shapers, 19 Apr. 2021, citychangers.org/walking-facts/. Accessed 27 July 2023.

Public transportation like trams and trains will make longer distances or commutes easier to traverse. Urban sprawl should be minimized and clustered around public transportation stops (see Part 3: "Opportunities for Expansion.") Physical activities of various kinds should be encouraged city-wide; consider Parks and Recreation organized programs for activities like jogging, rock climbing, yoga, martial arts, and certain sports.

Sports

Many popular sports like baseball, football and football (soccer) are extremely space-intensive and dedicated facilities like stadiums shouldn't be included in Augury. Green spaces can of course include fields (but planted with something other than grass, which is resource-intensive and wasteful[197]; moss is an effective ground cover and doesn't require fertilizing or watering once established[198]) for casual games, but we should avoid sports industries and large organizations like America has. The hyper-competitive culture surrounding professional sports encourages aggression, tribalism and prioritizes winning above physical and mental health, not to mention the laundry list of negative impacts on children.[199] In general, the industry is a massively unwarranted sink of resources, in overpaid athletes and coaches (who put debilitating pressure on players), hugely expensive facilities, and endless marketing. Boxing and wrestling especially should be discouraged for their intense toll on mental and physical health. We should not worship players of a game. Games are for fun—we should not glorify winning them in our society above the things that are really important. We should not

---

[197] Learn, Joshua Rapp. "Your Perfect Lawn Is Bad for the Environment. Here's What to Do Instead." Discover Magazine, 29 May 2021, www.discovermagazine.com/environment/your-perfect-lawn-is-bad-for-the-en vironment-heres-what-to-do-instead. Accessed 7 July 2023.
[198] Pino, Melissa. "Moss Lawn Care, Pros/Cons: Everything You Need to Know - 2023." Planet Natural, 16 Apr. 2023, www.planetnatural.com/moss-lawn/. Accessed 27 July 2023.
[199] Merkel, Donna. "Youth Sport: Positive and Negative Impact on Young Athletes." Open Access Journal of Sports Medicine, vol. 4, no. 4, 31 May 2013, p. 151, www.ncbi.nlm.nih.gov/pmc/articles/PMC3871410/, https://doi.org/10.2147/oajsm.s33556. Accessed 7 July 2023.

idolize athletes. We need a more casual, relaxed perspective on sports, our goal not to defeat our opponent and win at any cost, but to have fun and get a good workout. Sports are a pastime, entertainment—if players, watchers, or coaches are angry after or during a game, we're doing something wrong.

Consider some alternatives to serve as popular physical pastimes. Sports like volleyball, basketball, tennis, ping pong, badminton, and archery are less space intensive and don't require as large or dedicated facilities. Running and racing are flexible and can be organized anywhere around the city, and encourage good long-term cardiovascular health without being overly competitive. I highly recommend encouraging various martial arts, which instill values like respect, integrity, focus, perseverance, and restraint. Martial arts would be an excellent activity to include in schools as a form of Physical Education, as they would serve as a constructive outlet for strong feelings and a way for children to gain a better control of their bodies.

Sleep

Did you know that you'll die of sleep deprivation before starving to death? Though it's true we still don't fully understand sleep, we do understand how crucial it is for our health—and very few of us get enough. In his book, "Why We Sleep," Dr. Matthew Walker (a sleep specialist neuroscientist and psychologist), reveals the following:

> *"After thirty years of intensive research, we can now answer many of the questions posed earlier. The recycle rate of a human being is around sixteen hours. After sixteen hours of being awake, the brain begins to fail.* **[Adult] humans need more than seven hours of sleep each night to maintain cognitive performance. After ten days of just seven hours of sleep, the brain is as dysfunctional as it would be after going without sleep for twenty-four hours.** *Three full nights of recovery sleep (i.e., more nights than a*

weekend) are insufficient to restore performance back to normal levels after a week of short sleeping."[200]

Ask yourself the following: How often do you feel lethargic, sluggish, or foggy? Do you have trouble focusing or remembering, and do you find yourself stressed, impatient, anxious, or irritable? As you may have guessed, these are all symptoms of chronic sleep deprivation.[201] Our culture tends towards trying to get away with as little sleep as possible; we trade higher productivity for a shorter lifespan. In most cases, it's outside our control. A typical 9-5 schedule leaves little room for real rest. According to the Bureau of Labor Statistics, the average American works 8.8 hours in a 24-hour period (yet research suggests the average office worker is only productive for 3-4 of those hours).[202] Now subtract another hour for the commute—the average American daily commute is about half an hour one way.[203] Next, we'll subtract two hours for breakfast and dinner. Lastly, subtract around 8 hours for a good night's sleep, and we're left with around 4 hours for everything else. For a full-time high school student, that time is taken up for homework. For adults, you have only 4 hours per weekday to cook, do laundry, run errands, exercise, buy groceries, perform household maintenance, get dressed and undressed, bathe, fulfill your other responsibilities, and maybe, if you're lucky, have some time for socializing or self care . After the hustle and bustle of a full day, many of us are prone to what the Chinese call "revenge bedtime procrastination," in which we sacrifice sleep to have just a little more time to ourselves. Sometimes

[200] Walker, Matthew P. Why We Sleep : Unlocking the Power of Sleep and Dreams. New York, NY, Scribner, An Imprint Of Simon & Schuster, Inc, 2017.
[201] Suni, Eric. "Sleep Deprivation: Causes, Symptoms, & Treatment." Sleep Foundation, 3 Nov. 2022, www.sleepfoundation.org/sleep-deprivation. Accessed 21 May 2023.
[202] Curtin, Melanie. "In an 8-Hour Day, the Average Worker Is Productive for This Many Hours." Inc.com, Inc., 21 July 2016, www.inc.com/melanie-curtin/in-an-8-hour-day-the-average-worker-is-producti ve-for-this-many-hours.html. Accessed 21 May 2023.
[203] Bureau, US Census. "Census Bureau Estimates Show Average One-Way Travel Time to Work Rises." The United States Census Bureau, 18 Mar. 2021, www.census.gov/newsroom/press-releases/2021/one-way-travel-time-to-work -rises.html. Accessed 21 May 2023.

night is the only time where we're in control of our space and schedules—but staying up late doesn't mix well with a 9-5 schedule.

"Early to bed, early to rise" is our cultural norm. But for the estimated 30% of our population that are night owls, this principle doesn't ring true. Humans are not exclusively diurnal. There is a genuine, biological difference between people who are more active in the morning and people who are more active at night—in fact, according to research, there are as many as 6 different chronotypes for an individual's natural circadian rhythm.[204] This variation would have been advantageous in the era before modern lighting, but today, early rising is inexplicably considered morally superior. This is especially damaging to adolescents, whose circadian rhythms keep them up later and make it harder for them to be mentally active early in the morning.[205]

What if we embraced those differences? What if the culture of Augury facilitated an ongoing cycle—one group waking up while others were going to bed, all around the clock? Similar to the concept of a first/second/third shift, but without constant work. There would always be someone awake to run machines, handle crises, open shops—and no one would be forced into an arbitrary sleep schedule at the expense of their health. In the age of artificial lighting, why shouldn't we sleep when we want? Like hunger and thirst, why don't we just sleep when we're tired?

We need more than an average 8 hours of sleep per 24 hour period. Some nights we will stay up later, and some mornings we will sleep in—our sleep schedules don't have to be rigid, but overall, we just

[204] Putilov, Arcady A., et al. "Single-Item Chronotyping (SIC), a Method to Self-Assess Diurnal Types by Using 6 Simple Charts." Personality and Individual Differences, vol. 168, 1 Jan. 2021, p. 110353, www.sciencedirect.com/science/article/pii/S0191886920305444?via%3Dihub #bb0035, https://doi.org/10.1016/j.paid.2020.110353. Updated 13 Dec. 2020. Accessed 21 May 2023.

[205] Garey, Juliann. "Why Are Teenagers so Sleep-Deprived?" Child Mind Institute, 3 Feb. 2023, childmind.org/article/teenagers-sleep-deprived/. Accessed 21 May 2023.

need more sleep. The effects on our health and mood will fundamentally help us be better versions of ourselves.

Shelter

Surprised to see shelter listed among the fundamental needs for physical health? You shouldn't be—in survival situations, before finding water or food, the top priority is finding shelter[206]. Though often treated as a commodity, housing is considered a human right. In 1948, the United States signed the Universal Declaration of Human Rights (UDHR)[207], recognizing adequate housing as a component of the human right to an adequate standard of living. If we care about our people, we cannot tolerate letting them sleep on benches and doorsteps. Fortunately, exposure is less of a threat inside the walls of Augury, but let us consider the needs of every human being as equal to our own. Housing in Augury should not be operated for profit. Like water and air, it is a utility we all need to survive.

Options for housing in Augury will be more limited than on the surface. Hestia will be responsible for the construction of residential facilities of all kinds, including temporary accommodations like hotels and permanent lodgings in apartments—construction of any kind connected to the permanent structure of Augury should be performed by the department exclusively. Therefore, all residential space may have to be publicly owned (and maintained); a system to assign living space fairly will have to be designed. Things to consider include:

- Scaling up living space by family size
- Who gets relocated when newer facilities are constructed

---

[206] Hwang, Victor. "Four Needs of Wilderness Survival - Trackers Earth." Trackersearth.com, trackersearth.com/blog/four-needs-of-wilderness-survival/. Accessed 27 July 2023.
[207] United Nations. "Universal Declaration of Human Rights." United Nations, 2023, www.un.org/en/about-us/universal-declaration-of-human-rights. Accessed 27 July 2023.

- How residents can apply to relocate
- How residents can request specific locations (and ocean or atrium view)
- Who gets their requests fulfilled first; does anyone get preferential treatment? Are there raffles or lotteries? Is it first-come-first-serve?
- Do department heads get dedicated/special living quarters?
- How will the rental rate be calculated? Will government employees have this fee waived? Should all citizens?

Consider constructing all living units as an empty compartment (though insulated, wired for electricity, and equipped with water supply), with a set size depending on family size (600 ft^2 per person, 1200ft^2 for a family of 2, 2400ft^2 for a family of 4, etc.). The interior (including the floorplan/layout) could then be fully customized with modular building components like wall panels, electrical and plumbing fixtures, and flooring, all of which could be easily disassembled and reused in other units when residents relocate (consider using some kind of simulator to experiment with different layouts before construction). These fixtures would be purchased by the residents, so they have complete control over their living space. Consider perhaps offering a baseline "allowance" for fixtures that residents can use, and having to fund any further fixtures out of pocket. In general, occupants should have strong rights to their space while they occupy them—no surprise inspections, no trespassing or search and seizure by government officials without a legal warrant, and protection from surprise fees or rent increase. Residents should have contractually protected rights to their living space for their entire lives unless they decide to emigrate. Residents should be grandfathered into their living space for life so that in the event of a family member's death they would not be relocated against their will. Organizations like Home Owners Associations (HOAs) Should not be permitted; regulations for housing should be publicly established as a matter of legal policy and apply to all residents of Augury.

Old Age

In Augury, old age should not be considered derogatory. There are many facets to this cultural shift. Firstly, we must not idealize any particular phase of life—big life changes, new careers, new projects, self reinvention, expanding one's family, getting married, further education, and fun can happen in any phase of life. We have to get out of the way of thinking that we need to achieve "success" in our 20s, and in turn, we can't think of people older than 65 as if their life is already over. The other side of this comes with accountability for the elderly. With old age comes an increase in a neural quality called "thought crystallization," in which knowledge from previous experiences becomes stronger, but our ability to learn and accept new ideas (called "neuroplasticity") weakens[208]. Increased thought crystallization causes a person to tend towards stubbornness and xenophobia; preferring the old and becoming resistant to the new, and resistant towards learning and changing and growing. This means that older people are often dismissive, distrustful, or even fearful of change—a disadvantageous quality in the modern world. Many will dismiss antiquated mentalities (even problematic ones) by saying grandma or grandpa is a "product of their time." We must treat elderly people with respect and equality—which includes holding them to the same standard of accountability as we hold everyone else. Change is scary and uncomfortable, especially as one grows older, and they will need compassion, but that does not mean we should give up on them.

Growing older is hard. It has been since the dawn of humanity, and I can't imagine how much harder it is in this modern world, during the biggest technological boom in history. Too often we hide our elderly people away in nursing homes, isolated from the rest of the world, to keep them out of the way. If Augury is going to be a healthy community, we must take care of the members of our community at every stage of their life, whatever their needs, regardless of whether

[208] Cherry, Kendra. "What Are Fluid Intelligence and Crystallized Intelligence?" Verywell Mind, 23 Aug. 2021, www.verywellmind.com/fluid-intelligence-vs-crystallized-intelligence-2795004. Accessed 27 Aug. 2023.

they are able to work or not. We must care for our elderly, for their social needs and their physical needs. Augury should have something like a social security plan, where taxes pay towards a retirement account for all citizens. This fund, along with Augury's healthcare and housing arrangements, should mean that no one in Augury should have to worry about who will care for them as they grow older.

Overpopulation

To prevent overcrowding, population influx and capacity must be carefully balanced. The entire facility should always be designed with a *minimum* of 30% higher capacity than the current working population size to account for fluctuation, tourism, and refugees. Living space should include temporary space like hotels, permanent living spaces like apartments, and emergency space for temporary lodgings.

Overpopulation must also be controlled in regards to repopulation. Consider mandating application for parenthood, with required evaluation and counseling to ensure prospective parents are of sufficient physical, mental, and spiritual health to adequately raise children. This evaluation system should be designed by people like social workers who have experience working with children, and know what traits in parents can be particularly problematic. Multiple forms of birth control and sexual education should be freely and readily available.

Death

I have mentioned many times that space in Augury must be used prudently, and tragic as it may seem, this applies to the internment of the dead as well. We cannot justify the amount of space and energy it would require to store corpses indefinitely in a mausoleum (much less a cemetery) within Augury. That being said, we want to treat our dearly departed with respect and care, to tastefully memorialize their memory without preserving their corpse for eternity. When citizens die, I suggest one of the following should happen to their remains:

- The remains of the deceased are cremated then returned to the next of kin to be stored in an urn (at no expense).

- The remains of the deceased are cremated then turned into a composting pod for a choice of plants to grow from, which will then be planted somewhere in the city with a memorial plinth (at no expense).

- The remains of the deceased are cremated then turned into a custom 3D-printed reef starter for a choice of corals to grow from, which will then be installed somewhere outside the city with a memorial plinth (at no expense).

- The remains of the deceased are transported to the surface for burial elsewhere (at the expense of the next of kin).

Body Image

Vanity is a real threat to public health both mentally and physically, as unhealthy and unrealistic standards compel people (especially young people) to compromise various aspects of their health in the pursuit of cultural beauty standards and trends.

There are a few specific examples of unhealthy influences that we should avoid. Beauty pageants, especially for children[209], should be prohibited. The concept of a beauty pageant is inherently problematic, as it inevitably makes comparisons as judging some people as "more beautiful" than others, and one as the "most beautiful." Who are we to make a subjective judgment of beauty as though it is an objective standard rather than subjective, and inflict this standard onto the population as immutable law? This is not to mention the toll taken on the contestants to maintain the competitive

---

[209] Lovett, Megan. "The Dark Side of Child Pageantry." Wampus Cat Student News, 2 May 2022, wampuscatstudentnews.com/3360/opinion/the-dark-side-of-child-pageantry/. Accessed 12 Aug. 2023.

standards[210]. Modeling has similar issues, as showing an example of what beauty "should" look like in TV, advertisements, and print. Instead consider having businesses use 3D printed mannequins or 3D computer generated models[211] to display their product if needed.

Makeup is problematic—not inherently, but in the way we use it. According to New York dermatologist Dr. Michael Kurzman, "wearing too much makeup can expose the skin to chemicals and toxins...using a makeup brush too often without cleaning it will also build up dirt and bacteria that will clog your pores...makeup messes with our skin's natural cell renewal process...[and] might actually be prematurely aging your skin." Of course, we must not dictate people's choices for their own personal appearance (short of public modesty)—if a person wants to wear a little makeup, or a full face of makeup, or whimsical face paint, it is their choice. But if we do not constantly show the public perfectly airbrushed models, and if wearing a full face of makeup is not the cultural norm, hopefully people will not feel obligated to keep such unrealistic standards on their own skin. Digital retouching[212] and cosmetic plastic surgery should be prohibited. If citizens want these services, they can seek them elsewhere. We need to get used to what human beings look like, stop associating shame with how any feature of a person's body looks, and stop celebrating people for their looks. The only time that our concept of how attractive another person is to us is how we choose to present ourselves or when choosing a significant other.

Aesthetic beauty does not determine the worth of a person, and is largely a genetic roll of the dice. Our personal concept of beauty

[210] Sy, Michael P., et al. "The Dark Side of Occupation within the Context of Modern-Day Beauty Pageants." Work, 9 Jan. 2021, pp. 1–11, https://doi.org/10.3233/wor-205055.
[211] Barr, Sabrina. "Fashion Label Hanifa Puts on "Groundbreaking" Virtual Show Using 3D Models." The Independent, 25 May 2020, www.independent.co.uk/life-style/fashion/fashion-3d-show-runway-online-hanifa-instagram-live-congo-a9530551.html. Accessed 12 Aug. 2023.
[212] Ray, Amanda. "Picture Imperfect — Digital Image Manipulation Ethics." Artinstitutes.edu, 2015, www.artinstitutes.edu/about/blog/picture-imperfect-digital-image-manipulation-ethics. Accessed 12 Aug. 2023.

should be irrelevant to how we treat others. Our collective concept of true beauty in another person should be encouraged as not just superficial aesthetics, but more so a quality of a person's kindness, selflessness, and virtue. Our priority as a people should be health, which is so much deeper than the surface.

If the whole world were blind, how many people would you impress?
~Boonaa Mohammed

Ableism

One of the potential weak points I can foresee developing in the culture of Augury is an atmosphere of ableism. If we prioritize health above all else and consider it important to us, we may fall into the trap of conflating health with virtue, and seeing those of better health as superior to those with health related struggles. We must not ever let ourselves think that way. Health has a huge component of luck—one person's lackadaisical lifestyle might not show on them physically but would deteriorate the health of another. Every individual has different needs to achieve their personal best health, and personal best looks different for everyone. The sad truth of life is that some people are lifelong medical patients and will endure difficulty for their entire lives. Some of our friends and neighbors will struggle with chronic pain and illness every day and never get better even though they are doing their best. That is no reason to grow exasperated or give up on them. That is our reason to fortify our patience and support them one day at a time. Above all, we must remember that they are human beings, of equal value, who deserve as much respect as anyone else.

When it comes to mental and physical health disorders, the people with the conditions are the experts.[213] People with conditions should be in charge of designing systems that affect them—for example, have a person who uses a wheelchair at least weigh in on designing wheelchair accessibility throughout the city infrastructure, on whether

---

[213] Kennedy, Ian. "Patients Are Experts in Their Own Field." BMJ : British Medical Journal, vol. 326, no. 7402, 14 June 2003, pp. 1276–1277, www.ncbi.nlm.nih.gov/pmc/articles/PMC1126161/.

ramps are truly convenient or merely functional. Consult with people who are blind on where and how frequently to use Braille (consider a Braille sticker printer to use in public spaces). Ramps, automatic doors, elevators, ASL interpreters at events, and braille on public signs and menus should be the standard. Including accommodations like these makes the city more accessible to everyone. If the city is built for "universal design[214]," it will serve the entire community more effectively.

Sign Language

Sign language should be a national language in Augury and mandatory for all citizens to learn. The primary reason for this is communication in underwater environments—whether an individual ever intends to go diving or not, sign language will enable them to communicate with someone on the other side of an exterior window, and in extreme situations, the ability to communicate underwater may one day be a matter of life and death. Furthermore, vision-based communication proves useful in a variety of environments, such as very loud or very quiet environments, or over distance. Having the general population competent in sign language has the additional benefit of making Augury unusually welcoming to deaf communities.

---

[214] Universal design is the design of buildings, products or environments to make them accessible to people, regardless of age, disability or other factors. It addresses common barriers to participation by creating things that can be used by the maximum number of people possible.

B. Mental, Emotional, & Social Health

As previously stated, it's vitally crucial that we establish a hierarchy of health. Without a foundation of strong mental health, it will be harder to fight factors that threaten our spiritual health. We cannot adequately grow beyond selfish tendencies if those selfish tendencies are caused by our brains not being able to function beyond survival instincts. Helping others in any way is mentally taxing work—the stronger our mental health, the greater our capacity for compassion. Additionally, it may prove challenging to convince some groups of people that mental health is a reasonable concern—as with physical health, many suffer under the delusion that they are above therapy, or immune to trauma and chronic stress. As the author Robert Anton Wilson put it, "under the present brutal and primitive conditions on this planet, every person you meet should be regarded as one of the walking wounded. We have never seen a man or woman not slightly deranged by either anxiety or grief. We have never seen a totally sane human being." We must ensure that the health of the mind is considered with proper deference—as the brain is the most complex organ in the body, it is the most prone to malfunction, mistreatment, and disorder. Research supports "the notion that environments can affect, in one way or the other, people's mental health[215]." We must design that environment with care. Though mental illness appears to have become more culturally destigmatized in recent years, the greater effect has been what disability rights activist Marta Russel called "handicapitalism[216]," resulting in medications, self-care products, seminars, crystals, plants, and other superficial aids to mental health. The industry takes

[215] Helbich, Marco. "Mental Health and Environmental Exposures: An Editorial." International Journal of Environmental Research and Public Health, vol. 15, no. 10, 10 Oct. 2018, p. 2207, www.ncbi.nlm.nih.gov/pmc/articles/PMC6210156/, https://doi.org/10.3390/ijerph15102207. Accessed 21 July 2023.
[216] Adler-Bolton, Beatrice, and Artie Vierkant. "Capitalism & Disability: A Symposium on the Work of Marta Russell." LPE Project, 3 Oct. 2022, lpeproject.org/blog/capitalism-disability-a-symposium-on-the-work-of-marta-ru ssell/. Accessed 5 Aug. 2023.

advantage of vulnerable individuals without actually destigmatizing issues in a meaningful way. Our support of mental health must extend beyond the surface, and aim to grow rather than profit.

Before exploring the ways we can encourage a community of good mental health, let's explore the legitimacy of the issue by establishing a clear image of poor mental health in a community. Mental health is a complex, individualized subject, but we can discuss some generalizations to look out for. Stress should be our central concern for general mental health, for which mood and emotions are easy indicators. If people are impatient and quick-tempered, indulge in alcohol and other drugs, unenthusiastic and apathetic, anxious and depressed, or pessimistic, they are likely suffering from chronic stress. Stress and trauma left untreated overtime will grow into deeper issues. People suffering from unresolved trauma or mental health issues experience greater difficulty maintaining healthy relationships with other people.

Shorter Work Days

"The eight-hour workday is not based on the optimal number of hours a human can concentrate. In fact, it has almost nothing to do with the kind of work most people do now: Its origins lie in the Industrial Revolution, not the Information Age. In the late 18th century, 10-16 hour workdays were normal because factories "needed" to be run 24/7. When it became clear that such long days were both brutal and unsustainable, leaders like Welsh activist Robert Owen advocated for shorter work days. In 1817, his slogan became: "Eight hours labor, eight hours recreation, eight hours rest[217].'" Following WWII, productivity seemingly reached a peak, as the period from 1948 through 1973 is described as "The Golden Age of American Productivity." But while the labor productivity index (which represents the average annual real output per hour of office

---

[217] Curtin, Melanie. "In an 8-Hour Day, the Average Worker Is Productive for This Many Hours." Inc.com, Inc., 21 July 2016, www.inc.com/melanie-curtin/in-an-8-hour-day-the-average-worker-is-producti ve-for-this-many-hours.html. Accessed 21 May 2023.

workers) was only around 30 in the 1950s, it reached a whopping 107.79 in 2019, with minimal increase in wages.[218] We are three times as productive as we were in the 1950s, have far more powerful digital technology, yet we have far more workers struggling to survive and produce an enormous, unbelievable excess of our vital resources including housing[219], food[220], clean water[221], and energy[222]. The fact of the matter is that we don't have to work nearly as hard as we think we do to support ourselves with an adequate level of comfort. In fact, the average office worker is productive for less than 3 hours on an average work day[223].

We can produce all we need with just 4-6 hours on average per day, with a 4 day work week, instead of glamorizing "the grind" and our "work-hustle fetish culture[224]." Having more hours in the day gives us

[218] Sweet, Joni. "How U.S. Labor Productivity Has Changed since 1950." Stacker, 1 Sept. 2020, stacker.com/business-economy/how-us-labor-productivity-has-changed-1950. Accessed 18 July 2023.
[219] United Way NCA. "Vacant Homes vs. Homelessness in the U.S." United Way NCA, 28 Mar. 2023, unitedwaynca.org/blog/vacant-homes-vs-homelessness-by-city/. Accessed 18 July 2023.
[220] Shanker, Deena. "New Data Shows US Food Waste Is Getting Worse." Bloomberg.com, 20 Apr. 2023, www.bloomberg.com/news/articles/2023-04-20/the-us-has-a-food-waste-probl em-and-it-s-getting-worse. Accessed 18 July 2023.
[221] Perkins, Tom. "The Fight to Stop Nestlé from Taking America's Water to Sell in Plastic Bottles." The Guardian, 29 Oct. 2019, www.theguardian.com/environment/2019/oct/29/the-fight-over-water-how-nest le-dries-up-us-creeks-to-sell-water-in-plastic-bottles. Accessed 18 July 2023.
[222] Webber, Michael, and Joshua Rhodes. "The Solution to America's Energy Waste Problem Is Electrification." UT News, 19 Dec. 2017, news.utexas.edu/2017/12/19/the-solution-to-americas-energy-waste-problem. Accessed 18 July 2023.
[223] Vouchercloud. "How Many Productive Hours in a Work Day? Just 2 Hours, 23 Minutes..." Vouchercloud, 2019, www.vouchercloud.com/resources/office-worker-productivity. Accessed 18 July 2023.
[224] Kelly, Jack. "Finland Prime Minister's Aspirational Goal of a Six-Hour, Four-Day Workweek: Will It Ever Happen?" Forbes, 8 Jan. 2020, www.forbes.com/sites/jackkelly/2020/01/08/finlands-prime-ministers-aspiratio nal-goal-of-a-six-hour-four-day-workweek-will-this-ever-happen/?sh=3091c51f 3638. Accessed 18 July 2023.

more flexibility to care for our children, take care of errands, maintain our homes, spend time with loved ones, pursue passions and hobbies, rest and recharge, and engage in our culture and community. Some European countries have started pursuing shorter work days—amusingly enough, preindustrial medieval peasants had as much as eight weeks to half a year off and worked shorter hours per day[225]. In America, this may be a more difficult goal, as pursuing shorter work days and more free time would threaten industries who benefit from our willingness to pay for things we don't have time to do ourselves.

Humans are predators—some of the larger predators on earth. We often compare ourselves to bees or ants in judgment of our work ethic, imagining that every waking hour we spend working earns us a kind of virtue. This "busy bee" mentality is contrary to our biology; in fact, ants even spend a significant amount of time idle—not because they are lazy, but because idleness is an important part of survival. According to Dr. Daniel Charbonneau, who dedicated his Ph.D. thesis to studying ant idleness, "They really just sit there. And whenever they're doing anything other than doing nothing, they do chores around the nest, like a bit of brood care here or grooming another worker there..." From companies stocking supplies in warehouses to meet rising demand or employing contingent workers from external labor supply agencies, to computer systems performing better if equipped with reserve processing power, "the problem faced by all of these systems is how to optimally organize the supply or reserve workforce such as to minimize the costs of maintaining these reserves."[226] These ants are performing basic maintenance but otherwise biding their time, reserving resources, ready to act if a situation demands it.

---

[225] Schor, Juliet B. The Overworked American: The Unexpected Decline of Leisure. New York, Basicbooks, 1993.
[226] Stotle, Daniel . "Lazy Ants Make Themselves Useful in Unexpected Ways." ScienceDaily, 8 Sept. 2017, www.sciencedaily.com/releases/2017/09/170908205356.htm. Accessed 18 July 2023.

Importance of Idleness

Our idea of efficiency often leads us to ideas of filling our day with activities, making the most of our time and not wasting any of it. Most of the people I know operate on this model; I'm sure you know some people who are like this as well. Ask yourself this: do they seem a little tightly wound? Do they seem exhausted and overwhelmed, forgetful and impatient, maybe hard to reach? If we fill every spare moment we have with activity, it doesn't leave any room for anything else. If your schedule is full, you have no room for the unexpected or to be flexible. When a crisis does strike, you are left vulnerable without reserve energy and resources. I have known people like this to be non-receptive to new ideas or change, short-tempered, and hard to keep in contact with. There's no room in their heads for anything more—and life will always hand you more.

Now consider an individual who keeps a more leisurely pace. Imagine them to be calm and collected, easy-going, not easily ruffled. Their headspace is clear, their countenance serene. They are happy to stop and chat, interested in what you have to say, and emotionally equipped to stop and give their time to you if you need it. Between these two people I've described, which would you expect to be more prepared for the unexpected? Which has lower stress? Which has a higher capacity to help others? When we have more resources than we need and more free time, we have a cushion for times of struggle and more to share with others who have less.

There is a crucial importance for idleness and excess, and I can explain it with a simple metaphor: When pouring a cup of coffee, you never fill the cup all the way to the top. You always leave some extra room, so that if you bump the cup, you don't spill hot coffee and hurt yourself or someone else. We must leave ourselves extra room and extra time. Idleness and excess translate to flexibility and adaptability, which in turn translate to toughness[227] and resistance to stress. Working at a higher capacity may increase our output, but

---

[227] In materials science and metallurgy, toughness is the ability of a material to absorb energy and plastically deform without fracturing; the strength with which the material opposes rupture.

makes us brittle. Being overworked makes us careless and unobservant; more prone to mistakes. If we constantly try to achieve our best, that "constant" result is no longer our best, but our average—and if we give all we have, we have nothing left.

Social Health

"If a person cannot solve a conflict with a friend, how can they possibly contribute to larger efforts for peace? If we refuse to speak to a friend because we project our anxieties onto an email they wrote, how are we going to welcome refugees, immigrants, and the homeless into our communities? The values required for social repair are the same values required for personal repair."

— Sarah Schulman, *Conflict is Not Abuse: Overstating Harm, Community Responsibility, and the Duty of Repair*

In American culture, there are three topics considered inappropriate for polite conversation: politics, religion, and money. These are some of the most fundamental aspects of society: our rules and norms and the means by which we enforce them, our beliefs in morals and ethics and the values by which we interact with each other, and the means by which we provide for ourselves and exchange resources. Why do we avoid these topics? Do we think others do not have anything of value to share, that our perspectives are broad enough already? Is it because we don't want to be disagreed with, that we don't want our ideas to be challenged? Or is it because we don't know how to defend our ideas? Perhaps it's because we simply can't articulate what we believe.

Our culture struggles with healthy and fair communication. In our politics and our social media, we rely on sensationalism and logical fallacies to exchange information, which leads to misinformation, and excuses the general public from accountability to logic and fairness in debate. We don't know how to deal with conflict effectively so we avoid it, which makes us weak, irrational, ignorant, and ineffective at addressing the big problems of life. If we don't talk about our problems and talk through them we avoid trust and intimacy—which

lets tension build until it reaches a breaking point. This problem is evident in our highly divisive and polarized politics. This also makes it harder to form healthy relationships, because we don't know how to handle disagreements in a healthy way that pursues growth, rather than just fighting to feel like we've won.

To be unified and healthy, Augury needs to be a community of honesty. Our approach to disagreement must be selflessly working toward the goal of addressing the problem, rather than winning against an opponent. This principle should extend to the community as a whole, how we decide on policy, how we choose leaders, and even in our relationships with family, friends, and strangers.

The Arts

The government in Augury should be a great patron of the arts, and allocate consistent funding for beautification of public spaces in the city. Artwork like stoneworking, metalworking, sculptures, paintings, murals, and more should be commissioned and purchased from local artists and displayed in public spaces (along with credit to the artist); Hephaestus should maintain studios for artists to work in. The whole city should feel like walking through a museum. This should also extend to the practical arts and crafts, including furniture design, tapestry, jewelry, interior design, and so on.

Music, too, should be in public spaces—different spaces for performing should be built throughout the city, ranging from small alcoves to amphitheaters, equipped with seating, mic hookups, amps, and speakers (whatever equipment available, these spaces should work for both acoustic and electric instruments). This facilitation can extend to all the performing arts; the local theatre community should be encouraged with stages and auditoriums available for public use. Established performing groups should be able to register with the government and partner with Parks and Recreation to have their performances advertised and placed on public event calendars, and to facilitate ticket sales and patronage.

Writing and storytelling is one of the oldest forms of art; we can facilitate this by offering print-on-demand services through the library. A group of book-binding print-on-demand machines can produce a hard copy of any book written by a citizen to be kept in the library's collection; additional copies can be printed for purchase upon request or to meet market demand. The library can also host events like book signings, readings, and workshops with local authors. Consider the possibility of one or a few literary magazines produced through the library to facilitate short form or serial writings.

Video games are a composite art form; they combine storytelling and writing, performing arts, visual arts, and music. Augury should encourage video game development studios to settle in Augury and offer incentives. This will also bolster each specific art form, as game development studios may hire artists, musicians, and actors to provide their services. See Part 3: Economy, Video Gaming & Creative Industries for more detail on how this will be beneficial.

In short, the arts should be everywhere in Augury, easily found and saturating our community. With shorter work days, every citizen should be able to pursue some form of creative expression, whether they choose visual art, performing, arts and crafts, or whatever else. Being surrounded by so much creativity will in turn inspire us to be more creative.

> *"We don't read and write poetry because it's cute. We read and write poetry because we are members of the human race. And the human race is filled with passion. And medicine, law, business, engineering, these are noble pursuits and necessary to sustain life.*
>
> *But poetry, beauty, romance, love, these are what we stay alive for."*
>
> —Robin Williams as John Keating, Dead Poets Society, 1989

Library

The production of art is a worthwhile pursuit, but interpreting and appreciating art is also very important. "Art is inherently fueled by consumption. By making art accessible, the inspiration and analysis can continue in perpetuity[228]." A well-organized library facilitates and perpetuates art, creation, learning, and growth. With physical and digital storage, Augury can and should house one of the greatest collections of knowledge in history.

Modern libraries are not just repositories of books, but also academic resources, magazines, newspapers, ebooks, audiobooks, legal help, human connection, crafts, community events and resources, reference, artwork, music, video, and more. Augury's library should be a collection and center of human creation of all forms, easily accessible and navigable to all. The Great Library of Alexandria was part of a greater research institute to the arts called the Mouseion; we should consider something similar in Augury. One day, we might be glad we stockpiled our culture away safely, in the event of a cataclysmic disaster.

In addition to a collection of works and knowledge, libraries are centers of communities; hosting events for education, activities, and socialization. Ours should be as well—a place where people can exist without expectation of payment.

Education

The average student in America today endures greater stress than the average asylum psychiatric patient did in the early 1950s.[229] Here's a fact I feel no need to cite a source for: people's ability to learn is reduced when they are stressed. I currently work as a high school teacher, and less than a decade ago I was a high school student, so I can share some insight on the atmosphere from my personal experience: our education system is misery. The amount of

---

[228] Dais Johnston, 2023.
[229] Psychologist Dr. Robert Leahy.

pressure to perform and plan for their future that is placed on adolescents is trauma-inducing, driving many of them to chronic anxiety and depression, and, in some cases, even suicide. Everyone can tell that the only thing that matters to the administrative "powers that be" are test scores, but they suffer under the lurking knowledge that test scores and the almighty GPA can control the outcome of their entire lives, which inescapably influences their sense of self-worth.

I love to learn. I consider myself a polymath—a student of a wide variety of fields—I am curious by nature, and at the ripe age of 26, have been blessed enough to accumulate a book collection spanning fiction and nonfiction genres in the hundreds. I read Wikipedia and encyclopedias for fun, just for the sheer joy of learning. I have authored an 86,000 word novel and a 30,000 word novella, not including this nonfiction manifesto with over 300 cited sources. I hated school and I couldn't wait to get out of it. I got "senioritis" in my junior year of high school which didn't stop until I graduated college. Most students I meet feel the same way. The curiosity and joy of learning is beaten out of them by the time they reach puberty. There's something systematically wrong with the way we make children learn.

I can hardly overstate the importance of good education. It's an investment into our future as a race and a community. But I think there's a deep misunderstanding in what good education is. We have kids sacrificing their physical and mental health in some of the most formative years of their lives for knowledge they retain only long enough to pass a test. They graduate with little practical knowledge, little wisdom, little understanding of how to be a balanced, successful adult.

So, how do we do education differently? For starters, **less work**. The average American high school student spends around 2-3 hours per weeknight on homework in addition to their 8-9 hours spent in

school.[230] If they take advanced level classes to get ahead, that number can increase even higher. In order to retain the information we learn, we have to have more time to rest in between. If we just cram as much new information we can into our heads in a day, the brain doesn't have time to form new neural pathways. Our brains need time, repetition, and practical application to form strong connections and deep understanding and knowledge.

Secondly, we need to change *what* we teach our children. The general structure for American education was designed hundreds of years ago, back when literacy was uncommon and children didn't have access to the internet, and hasn't had a sufficient overhaul since. Kids (especially older kids), in my opinion, need applicability—they need to see how the things they are learning have real, practical value, and then they need the chance to apply it themselves. They need to put their hands to what they are learning and create something real. They need less paper and pencils and tests and more real world application, less simulation and technology and screens and more . Math, for example: a crucial skill, but in a vacuum it becomes abstract and boring. High school students study calculus—a specialized field of mathematics that few of them will ever use outside the classroom—but rarely study practical applications of math like personal finance or basic measurement and engineering. Why is calculus and precalculus a standard math class for high school students, but accounting, which is far more likely to be useful to them, an optional elective?

I propose a more holistic approach, and a recompartmentalization of classroom subjects into tracks or schools of study which combine related subjects into a balance of theory and application. For younger children, until around adolescence, the following tracks:

---

[230] Strauss, Valerie. "Does Homework Work When Kids Are Learning All Day at Home?" Washington Post, 1 Sept. 2020, www.washingtonpost.com/education/2020/09/01/does-homework-work-when-kids-are-learning-all-day-home/. Accessed 3 June 2023.

- **Indicina**
  - Programming
  - Computer Science
  - Mathematics
  - Finance
  - Data Science

- **Creo**
  - Engineering
  - Chemistry
  - Electricity
  - Electronics
  - Cooking

- **Intelligo**
  - Communication
  - Logic/Debate
  - Language Arts
  - Psychology

- **Nutrire**
  - Gardening/Horticulture
  - Animal Care
  - Medicine/First Aid
  - Biology

- **Vigeo**
  - Fine Arts
  - History
  - Social Studies
  - Sociology

Younger students would follow a division of 5 different tracks of subject compartments, focusing on only two to three tracks per day, to give them a general education and foundation for understanding the world around them and how certain subjects of study connect to each other. Gamification and application will help younger students to grasp more advanced concepts.

When they reach a certain age, they move on to secondary schools of study. To maintain a focus on the practical application of skills, subjects of study for older students are compartmentalized based on the **Executive** and **Judicial** departments. Students start by studying a little bit from each subject, then gravitate towards one or two tracks as they grow older and find proficiency and passion until they are eventually able to continue their education in the field itself through a paid internship. The corresponding departments work closely with students to connect their studies with the real world.

- **School of Bia**
  - Electrical Engineering
  - Electrical Contracting
  - Mechanical Engineering
  - Energy Physics
  - Nuclear Physics

- **School of Poseidon**
  - Plumbing
  - Plumbing Engineering
  - Fluid Dynamics
  - Chemistry
  - Hydrology
  - Marine Biology

- **School of Aether**
  - H.V.A.C. Engineering
  - Chemistry
  - Fluid Dynamics
  - Thermodynamics
  - Mechanical Engineering

- **School of Hephaestus**
  - Research & Development
  - Manufacturing Engineering
  - Mechatronics
  - Mechanical Engineering
  - Materials Engineering
  - Metallurgy
  - Chemistry
  - Sanitation
  - Waste Management
  - 3D Design

- **School of Hestia**
  - Structural Engineering
  - Architecture
  - Civil Engineering
  - Surveying
  - Welding
  - Construction Management

- **School of Demeter**
  - Horticulture
  - Botany
  - Biology
  - Biochemistry
  - Chemistry
  - Agriculture
  - Veterinary Science
  - Marine Biology

- **School of Apollo**
  - Network Engineering
  - Network Security
  - Library Science
  - Data Science
  - Data Analysis
  - Low Voltage Technology
  - History
  - Economics

- **School of Hermes**
  - Traffic Engineering
  - Civil Engineering
  - Mechanical Engineering
  - Fabrication
  - Mechatronics
  - Research & Development

- Aeronautic Engineering
- Marine Engineering

- **School of Minerva**
  - Programming
  - Computer Science
  - System Architecture
  - Software Engineering
  - Software Development
  - Machine Learning
  - Data Science
  - Security Engineering
  - Mechatronics

- **School of Aegle**
  - Medical Science
  - Nursing
  - Biochemistry
  - Pharmacology
  - Medical Technology
  - Physical Therapy

- **School of Physicality**
  - Public Health
  - Medical Science
  - Food Science

- Nutrition
- Exercise Science
- Physiology
- Community Health
- Preventative Medicine

- **School of Mentality**
  - Sociology
  - Social Work
  - Psychology
  - Counseling
  - Human Development
  - Family Studies
  - Community Development
  - Fine Arts

- **School of Spirituality**
  - Philosophy
  - Religion
  - Ministry
  - Social Work
  - Sociology
  - Ethics
  - Philanthropy
  - Leadership

One of the things which will contribute strongly to people receiving a strong education is not isolating them by age. In any given field of work or study, workers at all levels of experience and age should be able to interact. Learning from people older than us at more advanced stages in life and proficiency helps us grow and plan for our future more than only staying in a class of people our own age; the opportunity to teach people younger than us helps us grow ourselves and invest into our communities. This applies to allowing

older students to work alongside younger students as well as students working alongside adult professionals. The newest student and the heads of departments should not be in isolation from each other. In allowing them to mingle, we also allow new ideas, fresh perspectives, experience, and wisdom to mingle.

Culture Shifts

Apathy in our culture is a big problem. So much of our modern culture has evolved over time organically, without intention towards consciously designing our culture to the best of our ability. Are we truly satisfied to dismissively say "that's just the way things are?" We have to learn to care again, and when we do, we have to be sure that we care about the right things. There are certain norms and values, described in the sections below, that should be established in Augury to make it function better as a community.

Sensory Pollution

In an enclosed environment, sensory stimuli can grow overwhelming more easily. Air must be kept clean, free of overwhelming odors, but not sterile—consider introducing phytoncides and other chemicals found in forest air which improve mental health.[231] Sound ordinances must be enforced against sound pollution—everyone has the right to a reasonable degree of quiet in public spaces, so things like noisy neighbors and playing music out loud or talking on speakerphone on public transit must be held accountable. Industrial areas with heavy machinery should be kept away from residential and green spaces. Remember that water is an excellent conductor of sound. Brightly lit and flashy signs should be moderated. Dense crowding should be discouraged (also a safety concern).

Light pollution should be regulated, as use of lighting has a direct effect on mental state. Full-spectrum lighting should be used only during the day and from high overhead; otherwise, warmer lighting at

---

[231] Department of Environmental Conservation. "Immerse Yourself in a Forest for Better Health - NYS Dept. Of Environmental Conservation." Ny.gov, 2012, www.dec.ny.gov/lands/90720.html. Accessed 18 July 2023.

or near eye level should be preferred.[232] Exposure to blue light from cool-spectrum lighting must be minimized, as it causes a number of adverse effects including increased cell deaths and mitochondrial stress.[233] Fluorescent lighting should be avoided, as they can induce headache, fatigue, dizziness, nausea, eye strain, eye fatigue, and increased sensitivity to light. Studies have shown that lighting perceived as too dark is associated with a low mood level—but also with lighting that is bright. "Lighting that is experienced as 'just right' is associated with the highest mood level," according to Dr. Carla Marie Manly.[234] Of course, this threshold is different from person to person—the safest option is to keep lighting in public spaces low to moderate, while offering some areas that are more brightly illuminated, and let the individual choose the environment they need. Remember, an individual can bring a light into a public area if they need it, but if a public area is too bright, it's difficult to escape. To encourage a healthy circadian rhythm, lights in green spaces should be coolest during the afternoon, and be warmer in the hours near sunrise and sunset, to reflect more natural light changes.

Therapy Animals

So that we may enjoy a variety of benefits, domesticated dogs and cats should be allowed to roam freely around the public spaces in Augury. The primary reason for this is companionship, as people will derive joy from seeing, playing with, and otherwise interacting with animals. Living in proximity to animals like dogs and cats comes with several health benefits, including reduced anxiety and depression and fortification of the immune system (especially from an early age).

[232] Colino, Stacey . "The Lighting in Your Home Could Be Affecting Your Mood." Washington Post, 11 Apr. 2023, www.washingtonpost.com/home/2023/04/11/lighting-mental-health-well-being/ . Accessed 18 July 2023.
[233] Tao, Jin-Xin, et al. "Mitochondria as Potential Targets and Initiators of the Blue Light Hazard to the Retina." Oxidative Medicine and Cellular Longevity, vol. 2019, 2019, p. 6435364, www.ncbi.nlm.nih.gov/pubmed/31531186, https://doi.org/10.1155/2019/6435364. Accessed 26 Oct. 2019.
[234] Stathis, Jaime. "One Major Health Effect of Working under Overhead Lights." The Healthy, 9 Mar. 2023, www.thehealthy.com/home/overhead-lighting-vs-low-lighting/. Accessed 18 July 2023.

Having animals around and encouraging play with them will encourage exercise. If these are trained therapy animals, they can also seek out and help people in need of attention. Cats will help control populations of vermin which may sneak into the city with cargo. However, remember that cats hunt for sport—some protection will need to be afforded to birds.

These animals can be trained before they are released into the city to ensure they are not a nuisance. Aside from general behavioral traits, they can be trained to use automated water fountains, food dispensers, and litter boxes distributed throughout the city. RFID tags on collars can prevent overuse and help keepers keep track of the habits of each animal. Tires and other objects can be recycled into beds, then distributed at appropriate locations throughout the city. Dogs and cats should be selectively neutered to prevent overpopulation. When they are allowed to breed, parents should be of different breeds to increase genetic diversity. Traditional "purity" breeding limits the gene pool and is highly detrimental to the genetic health of animals[235].

Palliative Legislation

Though it will be a difficult challenge, avoiding **palliative legislation** will be one of the keys to Augury's long-term success as a balanced community. "Palliative" is a healthcare term which refers to treatment that addresses symptoms but not the underlying cause of a condition. Making laws which address the superficial symptoms of an issue is easier, but often punishes behaviors or activities which are not inherently bad, leading to an unnecessary and unfair restriction of people's liberties. Take loitering for instance—is loitering an inherently problematic activity? Or perhaps it could be an indication of lack of public common space in the community? Or, perhaps the goal is to repel specific groups of people, perhaps those too poor to

[235] Maldarelli, Claire. "Although Purebred Dogs Can Be Best in Show, Are They Worst in Health?" Scientific American, 21 Feb. 2014, www.scientificamerican.com/article/although-purebred-dogs-can-be-best-in-show-are-they-worst-in-health/. Accessed 20 Aug. 2023.

have access to or be welcome in a better public space. Alternatively, the goal may be to discourage criminal activity in a given area, in which case the criminal activity itself should be addressed, at its root. Palliative legislation addresses just the surface level of the problem and takes the shortcut, the easy way out. It punishes everyone to spite the few; instead of making fair rules, the rules are made equally unfair. It's a lazy way to address a problem temporarily and should be avoided in Augury. Communities of people are complicated, and so are the ways we address their problems, but if we give in to doing things the easy way, we let a few bad people ruin something for the majority. I will not be so lazy. It will not be easy to shift our mindset to see deeper than the surface when it comes to solving problems. It will take time. Often, the key to fixing these problems is investing in our communities on the front end rather than being reactive to behaviors we want to discourage—"an ounce of prevention is worth a pound of cure." This should not only be our approach to conflict in legislation, but in all the social issues we seek to solve.

Commodification

In America, almost everything we do is influenced by business and capitalism. Advertisements and clickbait and scams are inescapable; all-pervasive. Every year, I find out more and more cultural norms like expensive diamond engagement rings[236] and Christopher Columbus "discovering America" (rather than exploiting the Taíno)[237] are the result of marketing campaigns. Corporations buy politicians and lobby the government. Whether something is allowed to exist or not often comes down to whether or not it is profitable. Banks own everything. Forgive my callousness but allow me to be frank: I'm sick of it. I'm sick of everything in our culture coming down to money. I'm

---

[236] Uri Friedman. "How an Ad Campaign Invented the Diamond Engagement Ring." The Atlantic, The Atlantic, 13 Feb. 2015, www.theatlantic.com/international/archive/2015/02/how-an-ad-campaign-invented-the-diamond-engagement-ring/385376/. Accessed 27 Aug. 2023.
[237] History.com Editors. "Christopher Columbus." HISTORY, A&E Television Networks, 9 Nov. 2009, www.history.com/topics/exploration/christopher-columbus. Accessed 27 Aug. 2023.

sick of brands being everywhere and on everything, and every brand is eventually bought up by a bigger brand so I can't buy anything without doing business with a megacorporation. I'm sick of everyone having a side hustle and everyone on social media trying to sell me something because it's the only way they know how to survive. I'm sick of wealth defining whether you succeed in life or not. I'm sick of living in a corporatocracy. I'm sick of my concept of success being defined by how well I can make a sale or turn a profit.

This seems an opportune moment to define some misunderstood terms (since they are commonly misused). **Capitalism** is an economic and political system in which a country's means of production, trade, and industry are controlled by private owners for profit. As we have seen in America, this system is problematic because eventually some private owners accumulate more control of trade and industry (often by any means necessary, which develops into our trend of **vulture capitalism**), which they use to accumulate more control, and so on—without regulation, resulting in extreme economic inequality. "Capitalist culture promotes the accumulation of material wealth and the sale of commodities, where individuals are primarily defined by their relationship to business and the market[238]." A healthy economic system has citizens falling along a bell curve: the majority of the population will not make much more or much less than the median income. America is witnessing the rapid degradation of the middle class[239], and the extremely rich (top 0.1%) control over 15% of all wealth in America[240]. In the modern economy, so many gigantic international corporations like Amazon, Walmart, Nestle, Kraft, Coca Cola, Pepsico, Mars, and so on dominate the "free market" in such a way that the means of production, trade, and

[238] Wikipedia Contributors. "Culture of Capitalism." Wikipedia, 2 July 2021, en.wikipedia.org/wiki/Culture_of_capitalism. Accessed 26 Nov. 2023.
[239] Kochhar, Rakesh, and Stella Sechopoulos. "How the American Middle Class Has Changed in the Past Five Decades." Pew Research Center, 22 Apr. 2022, www.pewresearch.org/short-reads/2022/04/20/how-the-american-middle-class-has-changed-in-the-past-five-decades/. Accessed 26 Nov. 2023.
[240] Smith, Matthew, et al. "Top Wealth in America: New Estimates under Heterogeneous Returns." The Quarterly Journal of Economics, 29 Aug. 2022, https://doi.org/10.1093/qje/qjac033. Accessed 5 Sept. 2022.

industry are almost exclusively controlled by them in an **oligopoly**[241]. Anyone who looks at this evidence and believes capitalism is a healthy system for human societies has a far more favorable and permissive view than I do of the occasional predatory and greedy nature of humans . I believe the more vulnerable and less capable members of our communities should be protected from those who can and demonstrably will take advantage of them for personal gain.

Due to the nature of the executive departments and the need for public oversight to maintain a functional underwater habitat, the largest industries of Augury's governing system will bear resemblance (perhaps frighteningly) to **socialism**: a system in which the means of production, distribution, and exchange are **owned or regulated by the community as a whole**. This is not to say small, independent businesses cannot exist; they will be welcome. Any individual will be free to pursue wealth by means of their own labor, ingenuity, creativity, and skill—just not by means which exploit others. Everyone will pay their taxes, regardless of income or status. Augury should not house big corporations, franchises, or chains. We should put people and community first, and not pander to large businesses or people who worship love of money and profit above all else. Our actions and organizations, what we put our time and money towards, should reflect these values. Most people in today's world work at their jobs to further someone else's wealth; in Augury most of the population will work to build a community for the benefit of all. In America, being a public servant means enduring endless incompetence, inefficiency, and needless bureaucracy that's too big and ingrained to significantly change, but with a clean slate we can build an effective and efficient governing system from the start.

Addiction

Addiction of many forms is a rampant problem in modern society, but every form boils down to this: instant gratification. "When the brain has a pleasurable experience, a burst of dopamine…causes

---

[241] Faiola, Frank. "Oligopolies in America." Medium, 30 June 2023, medium.com/@faiola.frank/oligopolies-in-america-a3302537dfc8. Accessed 27 Nov. 2023.

changes in neural connectivity that make it easier to repeat the activity again and again without thinking about it, leading to the formation of habits. Just as drugs produce intense euphoria, they also produce much larger surges of dopamine, powerfully reinforcing the connection between consumption of the drug, the resulting pleasure, and all the external cues linked to the experience. Large surges of dopamine "teach" the brain to seek drugs at the expense of other, healthier goals and activities[242]." Some stimuli activate the reward centers in our brains more easily and immediately than others, which eventually forms a reliance—if we grow accustomed to a supply of dopamine from external stimuli, we soon are unable to function normally without it or tolerate more delayed sources of gratification.

Part of the problem is our culture of immediacy, which encourages us to pursue instant rather than delayed gratification; this also reduces our tolerance for delayed gratification. Social media actively and intentionally preys on this reward mechanism in our brains. Our phones give us social interaction, sate our curiosity, and relieve our boredom, all immediately, so much so that we grow irritable if we have to wait more than a few seconds for something to load. Around half of us admit to being addicted to our phones[243], and that number is increasing with the "iPad kid" generation. And while we're talking about children, let's spend a moment on sugar. Sugar consumption releases dopamine and opioids and is more addictive than cocaine, and forms dependency easily[244]. Sugar is added to almost all processed foods, and around 75% of Americans consume it in

[242] National Institute on Drug Abuse. "Drugs and the Brain." National Institute on Drug Abuse, July 2020, nida.nih.gov/publications/drugs-brains-behavior-science-addiction/drugs-brain . Accessed 7 Aug. 2023.
[243]Zauderer, Steven. "79 Cell Phone/Smartphone Addiction Statistics." Www.crossrivertherapy.com, 15 Dec. 2022, www.crossrivertherapy.com/research/cell-phone-addiction-statistics. Accessed 26 Aug. 2023.
[244] Avena, Nicole M., et al. "Evidence for Sugar Addiction: Behavioral and Neurochemical Effects of Intermittent, Excessive Sugar Intake." Neuroscience & Biobehavioral Reviews, vol. 32, no. 1, Jan. 2008, pp. 20–39, https://doi.org/10.1016/j.neubiorev.2007.04.019.

excess[245] (sugars should account for no more than 10 percent of your daily caloric intake), which can lead to anxiety and depression, bloating and diarrhea, cravings, muscle aches, headaches, fatigue, and nausea.

Pornography too is incredibly addicting and damaging—its abuse can deeply warp a person's perception of love, and is directly linked to increasing trends of sexual violence, abusive behavior, and human trafficking[246]. Young men in particular are affected by this—we are influenced from childhood that to be a man means sexual conquest, that the only acceptable form of love for us to accept is sexual gratification, and that the only loving physical touch acceptable to give or receive should be sexual in nature. Boys are starved for love and affection, and the only outlet they think they have is pornography. Of course, the other side is even darker—how horrible it must be to grow up as a girl in a world plagued by child pornography and sex trafficking, living in constant fear and being taught how to protect herself against predators. Every woman I have ever asked has been sexually abused or harassed at some point in her life, and sexualized by a man for the first time before she had even started puberty. Pornography as an industry must be illegal in Augury, as must offering or soliciting prostitution, escorting, and even sexual content in advertising, as well as every other industry which commodifies sex. We must relearn that sex is not a commodity or a skill or a service, but a very personal and private expression of love.

Of course, one thinks of the failure of alcohol Prohibition in America from 1920 to 1933, and the underground network of smuggling and speakeasies that sprang from it. Alcohol, along with tobacco, pornography, gambling, sugar, caffeine, and more are all addictive, but merely banning and criminalizing them would just make people seek them out illegally, and resent the government for encroaching

---

[245] Murray, Krystina. "Sugar Addiction - Find Help Today." Addiction Center, 10 Mar. 2023, www.addictioncenter.com/drugs/sugar-addiction. Accessed 26 Aug. 2023.

[246] Sharpe, Mary, and Darryl Mead. "Problematic Pornography Use: Legal and Health Policy Considerations." Current Addiction Reports, vol. 8, 9 Sept. 2021, https://doi.org/10.1007/s40429-021-00390-8.

on personal freedoms. How do we convince people to give them up? Some things, like alcohol or gambling, are not problematic in moderation. Others, like pornography and smoking (for air purity) must be outright banned. But for the rest of them, we must consider our living environment.

In the 1970s, American psychologist Dr. Bruce Alexander performed what's now known as the "Rat Park" experiment. In short, Dr. Alexander kept lab rats in two different environments: in one, the rats were isolated and understimulated, but provided with a choice between plain water and water laced with drugs like heroin or cocaine. In the other, the rats were kept in entertaining "rat parks," "where they were among others and free to roam and play, and to socialize. And they were given the same access to the same two types of drug laced bottles. When inhabiting a "rat park," they remarkably preferred the plain water. Even when they did imbibe from the drug-filled bottle, they did so intermittently, not obsessively, and never overdosed. A social community beat the power of drugs[247]." The takeaway from this experiment is the significance of community and environment in how we approach addiction. In short, addiction is a disease: our approach should be to heal, not to punish. People don't want to be addicted. If someone is so desperate to "take the edge off," we should explore what in their lives is tormenting them so much that they would turn to a substance or activity which could destroy them. Furthermore, industries and individuals who take advantage of others in this way must be held accountable.

Addictions start very often with a desire to self-medicate: a way to dull the pain or take the edge off life. If we make life easier, fewer people may feel the need to risk addiction. With therapy, relationships, and improved living conditions, maybe they can find healthier ways to dull pain.

---

[247] Sederer, Lloyd . "What Does "Rat Park" Teach Us about Addiction?" Psychiatrictimes.com, 2020, www.psychiatrictimes.com/view/what-does-rat-park-teach-us-about-addiction. Accessed 7 Aug. 2023.

Equality

America was first founded in an atmosphere of very selective freedom and equality—which was not, as many have suggested, just a product of their time. In response to the assertion "we hold these truths to be self-evident, that all men are created equal" in the U.S. Declaration of Independence, British abolitionist Thomas Day responded, "If there be an object truly ridiculous in nature, it is an American patriot, signing resolutions of independency with the one hand, and with the other brandishing a whip over his affrighted slaves[248]." From the beginning, our culture has been influenced foundationally by the genocide and systematic displacement of the native First Nations to steal their land[249], the enslavement and kidnapping of tens of millions of African natives[250], and discrimination, oppression, and xenophobia towards Irish, Italian, Polish, Hispanic, Middle Eastern, Asian, Pacific Islander, and Jewish people in the form of everything from naturalization laws, to discrimination, to internment camps[251]. This is not to mention deep misogyny in our culture, as women were deprived of the right to vote until the 1900s (August 26, 1920 for white women with the 19th amendment, but women of color couldn't fully vote until the Snyder Act in 1924, the Immigration and Nationality Act of 1952, the Voting Rights Act of 1965, and an extension to the Voting Rights Act in 1975, less than half a century ago[252]). We have spent the past 200

[248] Armitage, David. The Declaration Of Independence: A Global History. 76–77. Cambridge, Massachusetts: Harvard University Press, 2007. ISBN 978-0-674-02282-9

[249] J. Weston Phippen. ""Kill Every Buffalo You Can! Every Buffalo Dead Is an Indian Gone."" The Atlantic, 13 May 2016, www.theatlantic.com/national/archive/2016/05/the-buffalo-killers/482349/. Accessed 19 July 2023.

[250] Hacker, J. David. "From "20. And Odd" to 10 Million: The Growth of the Slave Population in the United States." Slavery & Abolition, 13 May 2020, pp. 1–16, www.ncbi.nlm.nih.gov/pmc/articles/PMC7716878/, https://doi.org/10.1080/0144039x.2020.1755502. Accessed 19 July 2023.

[251] Wikipedia Contributors. "Racism in the United States." Wikipedia, Wikimedia Foundation, 17 Jan. 2019, en.wikipedia.org/wiki/Racism_in_the_United_States. Accessed 19 July 2023.

[252] PBS. "Not All Women Gained the Vote in 1920 | American Experience | PBS." Www.pbs.org, 6 July 2020,

years trying to claw away from our discriminatory roots, fooling ourselves into believing that "racism is over" because we learned about Martin Luther King Jr. in high school. And these problems aren't limited to America; almost every developed country in the world has colonization, discrimination, and atrocities in their history. I won't spend any more time explaining that modern culture still has a big problem with bigotry—if I have to convince you, Augury is not yet the place for you.

We must be tolerant of others who are different, and embrace the diversity which makes us stronger, while simultaneously rejecting intolerance. In 1945 (shortly after WWII), philosopher Karl Popper described a principle which he called "the Paradox of Tolerance:"

> *"Unlimited tolerance must lead to the disappearance of tolerance. If we extend unlimited tolerance even to those who are intolerant, if we are not prepared to defend a tolerant society against the onslaught of the intolerant, then the tolerant will be destroyed, and tolerance with them.—In this formulation, I do not imply, for instance, that we should always suppress the utterance of intolerant philosophies; as long as we can counter them by rational argument and keep them in check by public opinion, suppression would certainly be most unwise. But we should claim the right to suppress them if necessary **even by force**; for it may easily turn out that they are not prepared to meet us on the level of rational argument, but begin by denouncing all argument; they may forbid their followers to listen to rational argument, because it is deceptive, and teach them to answer arguments by the use of their fists or pistols. **We should therefore claim, in the name of tolerance, the right not to tolerate the intolerant. We should claim that any movement preaching intolerance places itself outside the law and we should consider incitement to intolerance and persecution as criminal,** in the same*

www.pbs.org/wgbh/americanexperience/features/vote-not-all-women-gained-right-to-vote-in-1920/. Accessed 19 July 2023.

way as we should consider incitement to murder, or to kidnapping, or to the revival of the slave trade, as criminal[253]."

Here is my stance on intolerance, which must always be Augury's stance: from the very beginning, we must extend our beliefs of equality in value to all people, regardless of race, religion, gender, or sexuality. Every individual must be afforded the unalienable right to embrace their personal sense of identity without fear of discrimination or oppression. Inevitably, some individuals will disagree with the lifestyles of others. I do not begrudge them this, and I acknowledge their right to disagree with another's life choices. For example, I am a practicing Christian—I do not agree with many of the religious beliefs of others. But as the apostle Paul says in 1 Corinthians 5:12, It is not our responsibility to judge nonbelievers; only to judge fellow believers[254]. Disagreement is never justification for unkindness of any sort. If someone does not like another's lifestyle, if they wish to take part in a community of individuals, they should simply mind their own business. We each have the right to our own lifestyle as long as their lifestyle does not negatively affect those around us or the culture and society as a whole. We must be extremely wary of tribalism. **Extremist, nationalist, and supremacist organizations which glorify intolerance and discrimination such as the Ku Klux Klan and Nazism (or any symbols of those organizations) must be strictly forbidden and not even permitted a platform**. Augury is a place for people of all shapes, sizes, colors, ages, creeds, and origins to come together and be one. There is no room for bigotry.

I would be remiss if I didn't mention marital status in this section, as the LGBTQ+ community is such a significant point of political contention these days. Let me share my perspective: as I have mentioned before, America is not a theocracy, though it regularly

[253] Popper, Karl (2012) [1945]. The Open Society and Its Enemies. Routledge. p. 581. ISBN 9781136700323

[254] Apostle Paul. Holy Bible : Containing the Old and New Testaments : King James Version. New York, American Bible Society, 1611, 1st epistle to the Corinthians ch5 v12.

straddles the line. If it were, and forbade non-heterosexual marriage on that basis, it would have to prohibit divorce and remarriage by the same logic (it does not). I posit that government approval has no place in marital status. Marriage, divorce, and remarriage means different things for different people depending on religious/personal beliefs—to embrace true equality, I believe it should be a personal choice only. In Augury, consider a system in which an individual can choose a list of others to designate as their close family members, one a spouse/partner/heir/primary beneficiary, some as dependents, some as general beneficiaries, and so on. None of these people would not have to be genetically related. You choose your family, whatever it looks like—not the government.

Isolation

If you live in America, odds are you live in an urban or suburban attitude, along with over 83% of our population[255]—and if you live in one of these densely populated areas, you likely live in some kind of defined "subregion," such as a borough, a district, or a neighborhood—or maybe a single apartment complex. Think about this area you live in. How many people live in it? How many families? Is it dozens? Hundreds? However many people there are, there are some things that probably go along with each one. If you live in a suburban neighborhood, each house probably has at least one car. Each family most likely has their own washer and dryer, refrigerator and stove, and set of tools. Let's focus on a specific example: in the average American suburban neighborhood, how many power drills do you think there are? A power drill is a common and versatile tool, certainly an important staple of the standard toolbox—probably almost every family in that neighborhood has one, right? Now think about how often that drill gets used. Do most families use their power drill every day? Or does the power drill spend most of its time sitting in the garage or shed, waiting for its occasional use? In the same sense, most people who have their own pool will tell you they rarely use it. Some neighborhoods, on the other hand, have a communal

---

[255] University of Michigan. "U.S. Cities Factsheet." Center for Sustainable Systems, 2021, css.umich.edu/publications/factsheets/built-environment/us-cities-factsheet.

pool, which every resident in the neighborhood may come and use as they please—the cost of the pool is distributed and less expensive for each user, and the pool sees far greater utility. What if a neighborhood had a communal workshop and toolbox? How many power drills would be needed to satisfy the needs of the whole neighborhood? How much money could be saved by each family if they only had to collectively pay for a few power drills to share, rather than each buying one themselves?

The purpose of this mental exercise with the drill is to illustrate how our geographical proximity does not conflate with an efficient, cooperative community. This results in efficiency losses for individuals and families, as they have increased costs towards services and products that could be equally distributed in a community. Buying a drill for ourselves gives us greater independence with the drill we own, but is that level of independence meaningful enough to justify the cost? We still power the drill with electricity from the grid we all collectively pay for, on land and property we have to pay taxes on, subject to local ordinances and codes. The most independent lifestyle we could likely achieve is to have an off-grid self-sustaining homestead, which is a very difficult lifestyle because of how much work we have to do ourselves. Very few of us choose to live outside an established society, where we are subject to endless rules, regulations, costs, and fees. We suffer the disadvantages of living in geographical proximity to each other but not so many of the advantages. There are so many things necessary for modern life that would be more efficient and less expensive overall if we managed them cooperatively. So why do we each need to buy our own drill? Ask the people selling drills. The more separated we are, the less we rely on each other, and the more we have to rely on paid goods and services—the more willing we are to pay for greater convenience.

In America, the vast majority of us live close enough to see through our neighbors' windows, and yet we live in isolation. For many (if not most) of us, our norm is to keep to our immediate family and mind our own business—go to work, come home and spend some time with your spouse and children, then go to bed and do it all again. I

grew up in a neighborhood, lived in the same house for 15 years, and never even learned my next-door neighbor's name. As the world grows more dangerous, we become more withdrawn and wary of strangers. We rarely share resources or tools—nor time and burdens. Usually we only meet and grow close to strangers through the course of dating. Most of my generation has difficulty forming meaningful friendships as adults without the framework of school. All of these trends are of course not always the case, but they contribute to a growing trend of loneliness and isolation, which then self-perpetuates:

> "When people enter into an experience of loneliness, they trigger what psychologists call hypervigilance for social threat, a phenomenon Weiss first postulated back in the 1970s. In this state, which is entered into unknowingly, the individual tends to experience the world in increasingly negative terms, and to both expect and remember instances of rudeness, rejection and abrasion, giving them greater weight and prominence than other, more benign or friendly interactions. This creates, of course, a vicious circle, in which the lonely person grows increasingly more isolated, suspicious and withdrawn. And because the hypervigilance hasn't been consciously perceived, it's by no means easy to recognise, let alone correct, the bias. What this means is that the lonelier a person gets, the less adept they become at navigating social currents. Loneliness grows around them, like mold or fur, a prophylactic that inhibits contact, no matter how badly contact is desired. Loneliness is accretive, extending and perpetuating itself. Once it becomes impacted, it is by no means easy to dislodge. This is why I was suddenly so hyper-alert to criticism, and why I felt so perpetually exposed, hunching in on myself even as I walked anonymously through the streets, my flip-flops slapping on the ground."

> — Olivia Laing, The Lonely City: Adventures in the Art of Being Alone

We find endless ways to categorize and separate ourselves from others—by age, gender, ethnicity, religion, politics—and manage to isolate ourselves from people not like us. There's nothing wrong with spending time amongst kindred spirits, but if we avoid diversity, our perspective is narrowed. The pandemic illustrated better than ever how starved we are for meaningful connection and community. Consider the Young Adult fiction genre—so many examples are set in a setting like a boarding school, which is a convenient format for the close-knit, walkable community with friends we all desperately want.

Humans are social creatures, and deprivation of a healthy social community adversely affects our mental health. A human being *cannot* be healthy in isolation, and while prescription medication is often necessary to aid a person's health, it is no substitute for what we really need. We need meaningful relationships from people who will help us learn and grow—not just from people our age, or our religion, or our race. We have to learn to make and maintain relationships with people who are different, so we can love people who are different. To be a human is to endure pain and hardship, but we don't have to each pay for our struggles by ourselves and keep the pain locked behind fences and drawn curtains. Our problems become smaller when we bear each others' burdens.

"As iron sharpens iron, so one person sharpens another."
—Proverbs 27:17

Gratitude

We are enjoying a period of immense luxury and convenience, and there is so much we take for granted. "If you just did something like going to the supermarket and experienced it fully without the goggles of habit and categories you would go crazy with pure sense and joy[256]." How much happier would we be if we stopped now and then to merely appreciate all we have and all we can do, without comparison? In researching this topic I came across a video produced by Kurzgesagt — In a Nutshell, who address the topic very eloquently, thoroughly, and effectively:

> "Everybody is familiar with the feeling that things are not as they should be. That you're not successful enough, your relationship isn't satisfying enough, that you don't have the things you crave. A chronic dissatisfaction that makes you look outwards with envy and inwards with disappointment...In the last two decades, researchers have been starting to investigate how we can counteract these impulses...While gratitude may sound like another self-improvement trend, preached by people who use hashtags, what we currently know about it is based on a body of scientific work and studies.
>
> The predecessor of gratitude is probably reciprocity. When your brain recognizes that someone's done something nice for you, it reacts with gratitude to motivate you to repay them. This gratitude makes you care about others, and others care about you. This was important because, as human brains got better at reading emotions, selfish individuals were identified and shunned. It became an evolutionary advantage to play well with others and build lasting relationships.

---

[256] Tania [@boywaif]. "I had a French professor who once said if you just did something like going to the supermarket and experienced it fully without the goggles of habit and catégories you would go crazy with pure sense and joy. I think about it all the time. In a way this is all for him." Twitter, Dec 12, 2022, https://twitter.com/boywaif/status/1602301368705978369.

Scientists found that gratitude stimulates the pathways in your brain involved in feelings of reward, forming social bonds, and interpreting other's intentions. It also makes it easier to save and retrieve positive memories even more, gratitude directly counteracts negative feelings and traits, like envy and social comparison, narcissism, cynicism, and materialism. As a consequence, people who are grateful, no matter what for, tend to be happier and more satisfied. They have better relationships [and] an easier time making friends. They sleep better, tend to suffer less from depression, addiction, and burnout, and are better at dealing with traumatic events....In the best case, gratitude can trigger a feedback loop. Positive feelings lead to more prosocial behavior, which leads to more positive social experiences that cause more positive feelings...So, in a nutshell, gratitude refocuses your attention towards the good things you have, and the consequences of this shift are better feelings and more positive experiences.

The easiest gratitude exercise, with the most solid research behind it, is [acknowledging your gratitude]. It means sitting down for a few minutes, one to three times a week, and writing down five to ten things you're grateful for...In numerous studies, the participants reported more happiness and a higher general life satisfaction after doing this practice for a few weeks. And, even more, studies have found changes in brain activity some months after they ended. This research shows that your emotions are not fixed. In the end, how you experience life is a representation of what you believe about it. If you attack your core beliefs about yourself and your life, you can change your thoughts and feelings, which automatically changes your behavior. It's pretty mind-blowing that

something as simple as self-reflection can hack the pathways in our brain to fight dissatisfaction.[257]."

The takeaway from what we know about dissatisfaction is that it is relative. A person can be happy if they have very little while another person is dissatisfied with great wealth. The difference seems to lie heavily in our attitude—whether we choose to be happy with what we have, or constantly crave more. That deep, hollow emptiness that comes from dissatisfaction can lead to addiction and dangerous behavior as people take more risks to just feel something. Gratitude is not an alternative to professional counseling or medication for mental illness, but it will surely lend us a more optimistic outlook. The people of Augury will enjoy many luxuries and privileges—we must make sure we never take them for granted. We must appreciate that nothing we enjoy is guaranteed—from wealth, to comfort, to strength, to health—and might disappear tomorrow. We must use this attitude to keep ourselves humble and gratefully, and compassionate towards people who are less fortunate.

The Sickness of Greed

America and much of the rest of the developed world is characterized by competitive capitalism. Our people believe in survival of the fittest, of bootstrap philosophy, and it tricks us into thinking that rich people are winning and poor people are losing. We believe we are entitled to as much money as we can possibly get, and we crave as much profit as we can make. We idolize billionaires and scoff at those who struggle, and dream of the day when we too will achieve the American dream and be so rich that we can buy our way out of any problem we face, and ignore the struggles of the less fortunate. If I could identify one root cause of all the imbalance in the modern world, it would be love of money and greed. The extreme inequality of wealth has devastated the balance of our world, leaving billions to starve while fewer than 100 men and women control

---

[257] Kurzgesagt — In a Nutshell. "An Antidote to Dissatisfaction." Youtube, 8 Dec. 2019, www.youtube.com/watch?v=WPPPFqsECz0&ab_channel=Kurzgesagt%E2% 80%93InaNutshell. Accessed 11 Dec. 2022.

enough wealth to buy entire countries. The love of money is a terrible sickness, which inflicts a constant, selfish, undying hunger for more which can never be satisfied. The extremely powerful are greedy and gluttonous, consumed and blinded by this desire; they are beyond a sense of moderation or self discipline. If we allow it to continue in the human race unchecked, the few will decimate our limited resource supply. We must not allow those who are sick in this way to start accumulating power and resources. Instead, we should treat them as we would anyone with a destructive mental illness. We must all let go of our insatiable hunger for prestige, power, and privilege and instead hunger for compassion, kindness, and cooperation. We must have no love for money, but see it only as a tool, which we use for what really matters.

"I do not love the bright sword for its sharpness, nor the arrow for its swiftness, nor the warrior for his glory. I love only that which they defend."

— J.R.R. Tolkien, The Two Towers

C. Spiritual, Ethical, & Moral Health

"As our own species is in the process of proving, one cannot have superior science and inferior morals. The combination is unstable and self-destroying."
—Arthur C. Clarke

Oftentimes, people are too desperate with their physical or mental health to even think about spiritual things, which is why—even though I believe spiritual health is the most important kind of health—it's last on the hierarchy. Once we are fed, housed, well-rested, and we have inner peace, we have all we need to look to the needs of others.

Before exploring the ways we can encourage a community of good moral and ethical health, let's explore the legitimacy of the issue by establishing a clear image of poor spiritual health in a community. Poor spiritual health looks like selfishness. If our community has human beings struggling alone to survive, or industries which actively benefit by taking advantage of others, idolizes people who consider themselves better than others—if we have a weak or nonexistent sense of guilt and accountability for our actions or obligation to right our wrongs, we should consider it a moral failing. If the people of Augury are not kind, selfless, compassionate, and humble, we must immediately seek means of improving.

Altruism

Noun. *"Selfless regard for or devotion to the welfare of others[258]."*

Compared to mental and physical health, spiritual health may seem relatively unimportant—and more of a personal choice or journey than something a government department should be regulating. But since Augury has departments dedicated to promoting mental and physical health, spiritual health is crucial for balance.

---

[258] Merriam Webster, 2023.

Just like children, if we have our needs met and have to endure minimal struggle, we will grow spoiled. This leads to self-centered and immature behavior, entitlement, laziness, and eventually a kind of learned helplessness. Struggle to the point of mass suffering and chronic stress can kill us—but in the complete absence of struggle, we slip into hedonism. A person who lacks purpose and meaning will distract themselves with pleasure.[259] Without a culture of spiritual health, Augury will quickly fall apart. This cycle has repeated endlessly throughout history, as societies develop to the point of overcoming the struggle of survival through technology and support systems, then subsequently grow soft and complacent, and are eventually overcome by the hubris of their misconception of self-sufficiency. "Hard times create strong men. Strong men create good times. Good times create weak men. And, weak men create hard times."[260]

So how do we balance these? How do we design a community which protects its members from difficulty but does not deprive them of hardship altogether? It's simple: we seek out the struggle.

Augury will provide for our basic needs. Citizens will never have to worry about getting enough to eat, access to clean water, having a place to sleep, access to healthcare, or having clean air to breathe. They will have security in their ability to provide for themselves. That security establishes a strong foundation—which can be used to support a lackadaisical lifestyle, or to support greater courage. We see this phenomenon in children. Young children who cry and receive attention promptly grow braver, less sensitive, secure in the knowledge that they will receive support if they need it.[261] They are emboldened to pursue higher goals sooner. How we choose to use that security, that boldness, will define us.

---

[259] Viktor Frankl
[260] G. Michael Hopf, Those Who Remain.
[261] Powell, Alvin. "Children Need Attention and Reassurance, Harvard Researchers Say." Harvard Gazette, 9 Apr. 1998, news.harvard.edu/gazette/story/1998/04/children-need-touching-and-attention -harvard-researchers-say/. Accessed 27 Apr. 2023.

At the top of Maslow's pyramid is a section called "self actualization." His idea was that, if all the lower needs are fulfilled, we can begin to seek a sense of fulfillment and realizing our potential, becoming the best version of ourselves. This will look different for every person. For some, self-actualization looks like developing a skill or passion, like the arts; some creative contribution to the world. It may be a project, repairing a relationship, mastery of a skill, seeking answers or higher education, teaching others, or any number of things.

I posit that this stage of self-actualization is the stage at which we have the greatest capacity for selflessness. The foundation of having our needs met has the risk of enabling a sense of lethargy, lulling us into a sense of not needing or caring about others. But if we promote a sense of self-actualization, of becoming the best version of ourselves, and then think beyond ourselves, we can use that strong foundation to help others. We can care for the weakest members—not just of our own society, but distribute surplus supplies and humanitarian aid to people in need internationally. With extra time and energy, our basic needs met, and a surplus in our production, we can take in refugees of war or conflict, we can feed the hungry, care for the sick, rehabilitate the hurting. We must shift our culture away from the competitive, individualistic mentality and become generous—not begrudgingly, but be cheerful givers. We can do more. It is possible to help too much, as well—if we provide support to the point of enabling unhealthy behavior, or teaching learned helplessness. We must learn a proper balance of providing the right kind of support, driven by data and research, as each individual needs.

This is our purpose. The struggle of others will become our struggles and the world will be our neighbor. This is the key, I believe, to our spiritual health: to care for others as we care for ourselves.

"I slept and dreamt that life was joy. I awoke and saw that life was service. I acted and behold: service was joy. "

—Rabindranath Tagore

National Anthem

Good moral health and altruism should be the highest goal of Augury. To promote this, I propose the following song as our national anthem:

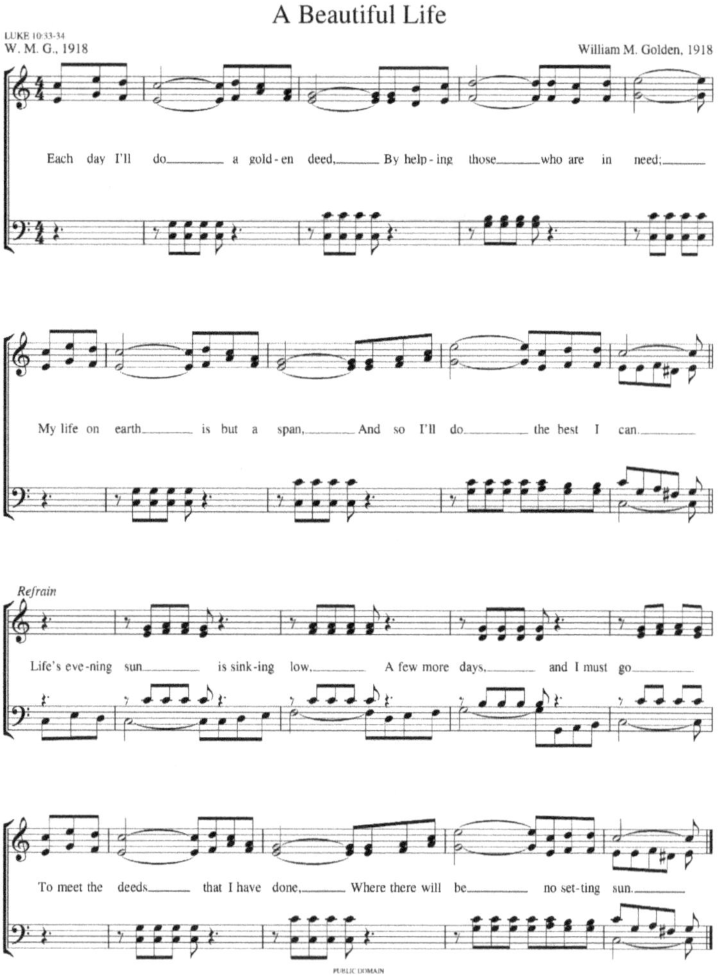

A Beautiful Life

LUKE 10:33-34
W. M. G., 1918

William M. Golden, 1918

Each day I'll do_____ a gold-en deed,_____ By help-ing those_____who are in need;_____

My life on earth_____ is but a span,_____ And so I'll do_____ the best I can._____

Refrain

Life's eve-ning sun_____ is sink-ing low,_____ A few more days,_____ and I must go_____

To meet the deeds_____ that I have done,_____ Where there will be_____ no set-ting sun._____

Religion

The first amendment to the U.S. constitution reads as follows:

> *"Congress* **shall make no law respecting an establishment of religion,** *or prohibiting the free exercise thereof; or abridging the freedom of speech, or of the press; or the right of the people peaceably to assemble, and to petition the Government for a redress of grievances."*

Most people remember the first amendment as merely a protection of free speech, and entirely neglect the very first mandate: "**no law respecting an establishment of religion.**" Many of the laws which shape our lives in America have no basis outside of Christian influences. America is described as "a Christian nation" (though this largely comes from propaganda campaigns during the Cold War[262]), and politicians use and abuse "Christian values" to manipulate the people. **Augury should not ever be a theocracy.** I am a Christian, and I do not want to create or live in a theocracy. I do not believe in theocracy. Faith is an individual choice, not a lifestyle to be enforced on others by anyone, least of all a governing body. Separation of church and state should be very carefully maintained. The Spirituality department, especially, **must never show preference to a specific institution or denomination of religion** (this is not limited to a specific religion, such as different denominations of Christianity, as someone I know has suggested as an interpretation of the American first amendment). The core value of the Spirituality department should simply be **impartial altruism.**

As far as the government interacts with religious organizations, here is what I propose. In America, churches are non-profit organizations and are exempt from some/all taxes, based on the assumption that they operate as a sort of charity organization. Though this concept

---

[262] Kirby, Dianne . "The Cold War and American Religion." Oxford Research Encyclopedia, 24 May 2017, oxfordre.com/religion/display/10.1093/acrefore/9780199340378.001.0001/acrefore-9780199340378-e-398;jsessionid=566E0E68832192B0C8DA972ECFD2DA1F. Accessed 18 July 2023.

often doesn't work in execution (when religious organizations take advantage of their tax exemption to maximize their wealth), I like the idea in principle. I propose that in Augury, we exchange facilities for service with religious organizations; perhaps something along the lines of a certain number of hours of volunteer charity service (for the Spirituality department, after which they may be delegated to another department) per registered member, which translates to a certain amount of square footage in a meeting place. This will ensure religious organizations operate as active and effective charity organizations in the community, and will prevent any organization from using wealth to be exempt from service. A simple, publicly available formula could keep the exchange rate fair and open. Any religious organization (with the exception of dangerously fanatical or radical organizations) which agrees to exchange service for a meeting place should be welcomed. If the service is repeatedly dissatisfactory or the organization is repeatedly disruptive or problematic to the rest of the community, the offer should be revoked.

Rehabilitation & Penal

Crime in Augury must be met with firm discipline. A functioning society is based on maintaining social contracts—when someone breaks the rules of cooperation, they break down a sense of safety and order, and reparations must be made to restore that. This is true of any society, but especially in an enclosed environment like Augury. Many people will confuse a lack of discipline with a sense of freedom, but make no mistake—if we fail to punish crime consistently and appropriately, Augury will not be a healthy community. Citizens must enjoy a sense of safety and security in Augury. They must not fear being attacked if they walk around alone (regardless of how they dress or their body language); nor should they fear being taken advantage of by someone else's neglect or mistreatment. But the conversation on our penal system can't stop there—what will define our approach to crime and justice is how we treat someone after they do something wrong. Do we endeavor only to punish them, to motivate them and onlookers by fear to never

break the law again? Or is our goal to rehabilitate them, to fix the core problem that compelled them to commit a crime in the first place?

The U.S. incarcerates a massive amount of people– around 0.7% of the population or 698 per every 100,000 people (which is about 20% of the world's population of prisoners, though America as a whole represents only 4.2% of the world population)[263]. America's incarceration rate is one of the highest in the world[264], and "approximately two-thirds will likely be rearrested within three years of release[265]." Our penal system is just that—a system focused on punishment far more so than rehabilitation, motivated to keep people in than to keep people out. This is because time spent inside prisons does little to improve the conditions of prisoners; poor nutrition, the high stress environment, overcrowding, abuse from guards, and solitary confinement all take an enormous toll on a person's health and leaves them worse off than when they went in, and ill-equipped to reenter society upon their release. Around 10% of American prisons are operated as private, for-profit entities and actively benefit from the retention of prisoners[266]. Conditions are cruel, inhumane, and destructive. Ex-cons are treated with prejudice from landlords, employers, and almost everyone around them.

---

[263] Wagner, Peter, and Wanda Bertram. ""What Percent of the U.S. Is Incarcerated?" (and Other Ways to Measure Mass Incarceration)." Prison Policy Initiative, 16 Jan. 2020, www.prisonpolicy.org/blog/2020/01/16/percent-incarcerated/. Accessed 25 July 2023.
[264] Carson, Ann. "Prisoners in 2021 – Statistical Tables." Bureau of Justice Statistics, Dec. 2022, bjs.ojp.gov/library/publications/prisoners-2021-statistical-tables Accessed 25 July 2023.
[265] United States Department of Justice. "USDOJ: FBCI: Prisoners and Prisoner Re-Entry." Justice.gov, 2019, www.justice.gov/archive/fbci/progmenu_reentry.html. Accessed 25 July 2023.
[266] Buday, Mackenzie, and Ashley Nellis. "Private Prisons in the United States." The Sentencing Project, 23 Aug. 2022, www.sentencingproject.org/reports/private-prisons-in-the-united-states/. Accessed 25 July 2023.

Consider in contrast the Scandinavian penal system, which has a reputation for focusing on rehabilitation. Prisoners are not treated like animals or sub-human. Depending on their facility, sentencing, and conduct, they may be afforded luxuries like flatscreen televisions, sound systems, their own personal clothes, wages, and even varying degrees of freedom, as some prisons don't even have walls[267]. Sweden, for example, has a prison population of about .07% or 74 per every 100,000 people[268], and only a 30% recidivism rate[269], which has been falling in recent years.

When a member of our community makes a mistake, we have a choice in how we want to respond. We can choose to punish them and make them suffer to motivate them through fear to never commit the crime again, or we can choose to help them and rehabilitate them so they are less likely to *want* to commit the crime again. Many principles in parenting apply to how we treat each other—the authoritarian parent focuses on the punitive, is strict and inflexible, is prone to demoralizing the child, ignoring their point of view, and not showing compassion, and may become abusive. Mistakes are met with harsh retaliation that breaks the child down. This focus on control and obedience results in an aggressive, socially inept, and rebellious child with low self esteem[270]. Alternatively, "authoritative parents are nurturing, responsive, and supportive, yet set firm limits for their children. They attempt to control children's behavior by explaining rules, discussing, and reasoning…children raised with this style tend to be friendly, energetic, cheerful, self-reliant,

[267] Larson, Doran. "Why Scandinavian Prisons Are Superior." The Atlantic, The Atlantic, 24 Sept. 2013, www.theatlantic.com/international/archive/2013/09/why-scandinavian-prisons-are-superior/279949/. Accessed 25 July 2023.
[268] World Prison Brief. "Sweden | World Prison Brief." Prisonstudies.org, 2017, www.prisonstudies.org/country/sweden. Accessed 25 July 2023.
[269] Holmgren, Martin . "Swedish Correctional Services: Growing Challenges in an Era of Change." JUSTICE TRENDS Magazine, 21 Mar. 2022, justice-trends.press/swedish-correctional-services-growing-challenges-in-an-era-of-change/. Accessed 25 July 2023.
[270] Trautner, Tracy. "Authoritarian Parenting Style." MSU Extension, Michigan State University, 19 Jan. 2017, www.canr.msu.edu/news/authoritarian_parenting_style. Accessed 25 July 2023.

self-controlled, curious, cooperative and achievement-oriented[271]." Our governing style should follow a similar model. If our goal is to reduce crime and the number of criminals, our approach must be centered around rehabilitation. To rehabilitate people, we have to be goal-oriented and address problems at their source. Either on an individual basis or as a societal trend, what motivates a person to commit a given crime? If the motivation is desperation, can the person's living conditions be improved? If the situation was circumstantial, how can the circumstances be avoided—or can counseling improve internal pressure or motivations?

FEDERAL OFFENDERS BY TYPE OF CRIME[1]
Fiscal Year 2018

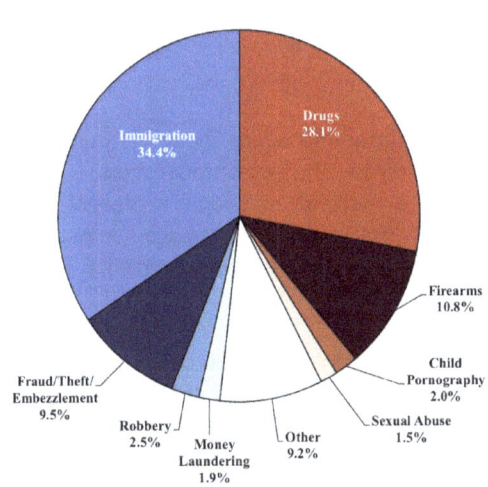

¹ This figure includes the 69,425 cases reported to the Commission. The Drugs category includes trafficking and simple possession. Descriptions of variables used in this figure are provided in Appendix A.

SOURCE: U.S. Sentencing Commission, 2018 Datafile, USSCFY18

---

[271] American Psychological Association. "Parenting Styles." Https://Www.apa.org, June 2017, www.apa.org/act/resources/fact-sheets/parenting-styles. Accessed 25 July 2023.

According to data provided by the United States Sentencing Commission, in 2018 more than half of federal crimes were either immigration and drug-related[272]—both of which speak to deeper issues than mere selfishness. In some cases, very violent or cruel crimes must be met with firm accountability to defend the victim(s). However, in many cases, criminal activity can be considered an indicator of some failure or shortcoming in culture. Should they still be punished for breaking the social construct of the community? Of course it should, and where possible, the punishment should be tailored to fit the crime (towards righting the wrong), like community service or fines (note that fines should always be a percentage based on income; otherwise fines will not be a meaningful deterrent to wealthier citizens). Rehabilitating people is always a combination of improving their physical, mental, and spiritual health and should be treated as such. The main goal should be that after a criminal has served their sentence, they are well equipped to reenter society, in better condition overall than before.

As in America, no citizen in Augury should be punished unless they are convicted of a crime by a jury of their peers. Every citizen must have the right to a fair trial, and legal representatives should be publicly employed exclusively and assigned impartially. Wealth or status should never give a person or organization an advantage in a court in the pursuit of truth.

We must also remember that penalization is more reactive than preventative—and an ounce of prevention is worth a pound of cure. Threat of punishment can be an effective deterrent against crime, but even more effective is strong support and social systems and a fair and balanced community.

---

[272] Pryor, William, et al. Sourcebook of Federal Sentencing Statistics United States Sentencing Commission. 2018.

Unforgivable Crimes

Certain extreme crimes like rape, aggressive molestation, molestation of vulnerable people (children, elderly, handicapped, etc.), human trafficking, physical torture, and first degree murder, especially serial repeat offenses of these crimes, must be punished drastically. Capital punishment should only be used as a last resort, as it could incentivize perpetrators to murder their victims to avoid getting caught. Banishment is appealing, but unfair to the nation the perpetrator ends up in. Chemical castration and/or permanent incarceration may be the most viable option. Though incarceration is not ideal, sometimes it's necessary in order to maintain a safe environment for the general population. Perpetrators of past sexual and violent crimes should not be allowed to emigrate to Augury.

Fortunately, the ocean is facilitative to a high-security penitentiary. Consider a facility in deeper water than the rest of Augury, where water pressure would make escape by swimming impossible. Supplies could be transported in an unmanned compartment—if this compartment traveled quickly enough, it would prevent stowaways by threat of decompression sickness (alternatively, the compartment could be divided into smaller compartments internally too small for a human to fit, or the compartment could be programmed to flood before returning). This facility could be self-sufficient and remotely operated through the use of mechanisms like cameras, microphones and speakers, shock collars, and so on, which may reduce or eliminate the need for internal guards (though guards would still need means to quickly enter and exit the facility when needed; consider perhaps having it only accessible from above to use gravity as an obstacle). Inmates should not be exposed to any provocative stimulus. Sedatives in food or water could reduce the risk of violence. This facility should not be completely miserable—inmates should enjoy reasonable comfort that they may be content to calmly live out the rest of their existence isolated from the rest of the world.

City Manager's Office

The City Manager's office answers directly to the Judicial branch and is responsible for coordinating with all of the departments in the Executive branch. The primary function of the department as a whole is coordination of the collective efforts of all other parts of the government; making all the different parts of the machines smoothly working together. A qualified candidate for City Manager has significant leadership, communication, and creative problem-solving skills, and must be familiar with a wide range of subjects to be able to unite a diverse group of workers. The main directives of the City Manager's office is to ensure that the Executive branch departments are fulfilling the city-wide priority of health, as defined by the Judicial Heads (and to enforce their guidance), and to help the departments cooperate in any way necessary, mediating disputes, aiding communication, and assisting optimization to every extent possible. To this end, they should perform regular check-ins on the departments to assess needs.

While some officials of the city manager's office have day-to-day tasks, most upper-level officials intentionally maintain an open schedule. This is crucial to ensuring that they are ready and open to addressing surprises, crises, and emergencies wherever they arise. In an underwater city, capable people need to be ready and able to drop what they're doing at a moment's notice and address the unexpected.

Part of the City Manager's responsibility is to maintain a connection of accessibility between Augury's people and government departments. This includes hearing citizens' concerns and addressing them, as well as keeping the people updated on what the various departments are working on or planning. Consider having the city manager appear on a regular news segment to give a general update on government projects, filmed in a public place like a cafe where passersby are free to sit and watch. Transparency is important; the activities of Augury's government should be common

knowledge to the citizens it answers to, just as the government's activities should be accountable to them.

Some of the most important functions of the City Manager's office is general management and organization, including **economics**, **tourism**, and **emergency response**. For such areas as these where protocol may be less a question of scientific optimization and more determined by popular preference, the City Manager's office should outline alternative policy plans in advance that are designed to shift policies towards attaining a specific priority without throwing Augury's entire governing system off-balance. For example: a few different versions of departmental budgeting plans could be prepared, with all the necessary adjustments planned out and the advantages and drawbacks analyzed in advance, so that if "plan A" is not working due to specific circumstances, "plan B" could be explored. Or, if more attention needs to be diverted to some income streams at the expense of others, plans could be devised—a "tourism priority" or "data center priority" plan. This method could be applied to insurance, retirement, taxation, and more.Of course, these premade plans would not have to be the only option available (and citizens should be able to submit their own plans), but could be an easy alternative to smooth transitions if a shift in priority was needed. Additionally, having pre-made plans designed to balance all necessary adjustments city-wide would make voting for change in policy easier.

Income Streams

No society that involves people and limited resources is without an economy. Additionally, though Augury is designed to be self-sustaining, it will still benefit from participation in the international economy, exporting resources and services that are easier to procure from an underwater vantage point, and importing exotic resources that cannot be locally manufactured. Augury's economy must be carefully monitored and regulated to prevent significant financial inequality among its citizens. The systems and executive departments are designed to use technology and engineering to

produce a surplus of critical resources. A significant financial surplus/windfall is also a must, to provide stability during times of crisis and to provide for those who can't provide for themselves. Taking care of people is expensive—but not as expensive as letting them suffer.

Data Center

The digital real estate industry grows more and more every year, and one of the biggest costs of maintaining servers is cooling. Ocean water is very cold and able to absorb a lot of heat—and as Microsoft has already found, "underwater data centers are reliable, practical, and use energy sustainably." With direct access to undersea communications cables, near limitless energy, and an abundance of cold water, Augury gets most of its national income from renting server space in its underwater data centers, run by Apollo. It's a stable, scalable source of income that reliably grows as the online economy grows.

Tourism

Who wouldn't want to visit an underwater city? After crucial facilities for walls, power, air, and water are set up, the first facilities built in Augury should be a full-service hotel, and later be expanded into a convention center. In time Augury will become a destination venue for conventions, exhibitions, expositions, trade shows, weddings, festivals, retreats, and more, with attractions and features found nowhere else in the world. Augury will sit 30-40 meters underwater. Modern materials like transparent wood composites, synthetic sapphire, and even transparent aluminum oxynitride ceramic (ALON)[273] can be made into windows strong enough to allow for spectacular, unimpeded views even on the ocean floor at a fraction of the thickness needed with acrylics or glass. Since coral grows freely on the exterior of the buildings and oxygen released by the mineral aggregation encourages ocean life, those views will be pretty

---

[273] Wikipedia Contributors. "Aluminium Oxynitride." Wikipedia, Wikimedia Foundation, 11 May 2019, en.wikipedia.org/wiki/Aluminium_oxynitride. Accessed 1 May 2023.

spectacular. With convenient access to the ocean floor, Augury will offer perfect conditions for tank, helmet, and even free diving. Visitors could explore vibrant reefs, colorful fish, and an octopus' garden dozens of meters below the surface. If they want to see the ocean but don't want to be in it, they could take a submarine tour to see larger formations or visit nearby shipwrecks. And don't forget the appeal of the ocean's surface, on which tourists could enjoy various water sports or relaxing on boats and floating islands.

In between events and attractions, visitors will be able to enjoy stretching their legs for a walk or sitting down for a picnic in any of Augury's spacious atriums. They will be able to eat at local restaurants with food grown in the city, browse shops stocked with goods crafted by local artisans, and attend regular events hosted by a welcoming, easy-going community. Augury will be a perfect place to slow down and smell the flowers.

Some visitors will take interest in the opportunity to take a tour of the facilities or lessons with any of the Executive departments to see how Augury works, and meet the workers who make it happen. They will be able to visit Bia's nuclear-Stirling generators and learn about thorium salts, help Demeter harvest fruits and vegetables from aquaponic arrays, recycle plastic into usable products with Hephaestus, or watch a technician service the underwater data centers in Apollo—perfect for STEM enthusiasts.

Tourism volume, of course, must be regulated—Augury will always have a hard maximum capacity, and this quantity of inhabitants must be given a wide berth in case of emergencies. Furthermore, it's important to ensure high tourism traffic is not detrimental to the exterior or interior ecosystem. With tourism comes litter.

Convention & Exhibition Center

Augury will be a hotspot of many industries: marine biology, sustainable technology, video game development—all of which organize events like conventions and exhibitions in which interested parties gather from all over the world. Imagine the possibilities of

Augury as that venue. The unique location would keep it in high demand, and the tourism would keep a steady stream of traffic, bringing in significant income. Organizers would take care of organizing and planning the event; we would only need to provide the space. Augury's convention center should include concert halls, lecture halls, meeting rooms, and conference rooms to accommodate events of all kinds. These facilities will also benefit citizens, by providing them with entertainment as well as a constant "revolving door" of new and interesting people, helping them to feel less isolated.

Retreat Center

Augury's prime directive of the prioritization of health can be combined with a tourism industry in a retreat center of sorts. Distinct from a hotel, a retreat center offers dormitory-style overnight accommodations but also offer meals, activities, meeting rooms, and chapel space (for religious organizations) and can be rented by religious, business, or social groups as a place to step back from everyday life and take time to "renew, reflect, and recenter one's life[274]." Geographical isolation and the serenity of the ocean could make Augury an enviable location for visitors to refresh themselves and bond, away from their daily routine. Retreat programs of various kinds could be offered, for example: religious or spiritual recentering, health and wellness, and team building. This retreat center could also function as a uniquely scenic wedding venue.

Consider including a spa and/or bathhouse to the retreat center, which could service the whole population in addition to tourists. Spas have a natural connection to the ocean—salts, seaweed, and other marine materials are often used in health treatments. Proximity to the ocean gives us easy access and authenticity, alongside soothing views of the ocean floor.

---

[274] "What Is a Retreat? - FindTheDivine Retreats Online." FindTheDivine, 13 Feb. 2018, findthedivine.com/retreats/. Accessed 25 July 2023.

Primary Induction Lighthouse

A surface lighthouse will be Augury's permanent link to the surface—through an escalator in an angled "bathyduct[275]" (to make the descent slower), visitors, tourists, citizens, air, and cargo will come and go, moderated by customs and security checkpoints inside the lighthouse. Visitors should be reminded by security staff of cultural expectations—for example, that Augury doesn't have the same approach to hospitality as some other countries—abuse of business or government workers will not be tolerated, and status as a customer or tourist does not absolve one of personal responsibility. Controlled substances like explosives, firearms, and internal combustion engines of any kind are strictly prohibited from entering the city, and weapons and produce should be controlled. Visitors suffering from a contagious illness should be asked to wait until they recover before entering the city unless they are in need of emergency medical attention.

For citizens of Augury, the lighthouse will be their link to the surface. They will be welcome to come and go as they please, of course, but would have to wait for the ferry to take them from the lighthouse to the nearest mainland, as Augury will be able to receive seaplanes but not have a typical airport. Augury will see more submarine traffic than most parts of the world, but unless someone privately owns one, they're not likely to hitch an underwater ride anywhere—yet.

On the underwater end, the lighthouse will be attached to the Postal Office. If a citizen orders a package from anywhere other than the nearest mainland, it'll be first shipped to a mainland post office, then transported over on the ferry, after which it'll be waiting for them in the local post office, as soon as it clears customs in the lighthouse and is sent below.

As the lighthouse is the only part of Augury above water, it can be outfitted with some solar arrays. Consider using colored solar panels

---

[275] Portmanteau of the prefix "bathy-", which refers to deep water (like in "bathysphere"), and "duct," which is a tube or passageway (like in aqueduct").

237

to make artful "stained glass" windows, which can function even without direct sunlight.[276]

Potential Points of Interest

Though Augury's location alone will make it a desirable tourism hotspot, the addition of specific points of interest will solidify the appeal even after the novelty of an underwater city has faded. The convention center will be a firm anchor for this, but I have included a few other additional attractions below:

Recreated Ancient Wonders

Consider points of interest inspired by the seven ancient wonders of the world:

- Great Pyramid Observatory
    - Inspired by the Great Pyramids of Giza, which have some significance associated with astronomy—as Augury will be out at sea, light pollution will be minimal so this facility could be dedicated as an observatory or planetarium.

- Hanging Gardens
    - Inspired by the Hanging Gardens of Babylon, which were originally intended to resemble a lush biome. Consider a tall, terraced structure with complex fountains inside a larger atrium to support large plants—just like the original, this version of the Hanging Gardens may provide a welcome refuge for citizens who miss the forest.

---

[276] Meinhold, Bridgette. "Colored Solar Panels Don't Need Direct Sunlight." Inhabitat - Green Design, Innovation, Architecture, Green Building | Green Design & Innovation for a Better World, 18 Sept. 2009, inhabitat.com/colored-solar-panels-dont-need-direct-sunlight/. Accessed 17 July 2023.

- Statue of Hephaestus
  - Inspired by Statue of Zeus; Statue of Hephaestus at his forge outside the department facility. Though no material is completely immune to biofouling, 3D-printing with a copper alloy will be resistant and thematically appropriate.

- Hospital of Artemis
  - Inspired by the Temple of Artemis—as Artemis is the ancient Greek goddess of nature, childbirth, and care of children, this facility could be dedicated as a world-class fertility and reproductive clinic.

- Mausoleum Reef
  - Inspired by the Mausoleum at Halicarnassus, a framework for an artificial memorial reef that will be built with the cremated remains of Augury citizens.

- Colossus of Gaia
  - Inspired by the Colossus Rhodes and Lady Liberty; a large statue of the ancient Greek goddess of the earth, Gaia, situated outside the city. This statue could be constructed of stone designed to encourage coral and aquatic plant growth, and eventually be overgrown with life.

- Lighthouse and Library (themed after Alexandria)
  - The primary induction lighthouse will be modeled after the Lighthouse of Alexandria, with a rectangular section under an octagonal section under a cylindrical section.

Biorhythmic Plant Clock

Since all the plants in Augury are going to be wired into a monitoring system anyway, why not use them to create beautiful music? Using a psychogalvanometer (the same technology in a polygraph machine), the micro-fluctuations of plants' biorhythms can be converted into musical tones. Consider using these tones as "church bells" which sound off on the hour throughout the city as an example of "local flavor."

National Holidays

The people of Augury will certainly celebrate popular holidays like Christmas, Hanukkah, Kwanzaa, Yule, Halloween, Dia de los Muertos, New Year and Chinese New Year, Ramadan and Eid al-Fitr, Diwali, and so on—but I believe Augury should also have holidays of its own. So, what should we celebrate? What do we celebrate? Here is what I propose: a holiday dedicated to each of the executive and judicial departments, in which everyone else except the members of that department perform that department's duties. This will help the population to see the value of the work each other does, and see how important the foundational labor of our community truly is. Garbage collection as an example springs to mind—many people in American culture look down on garbage collectors, but if they had to do their job for just one day a year, they'd likely appreciate those workers a lot more.

Of course, important landmarks in the history of the city should be celebrated as well. We (hopefully) won't have an independence day, but perhaps a "founder's day" or something similar. Holidays should be a day of rest for anyone that chooses to observe them rather than showing preference to one religion. We don't want to get carried away with days off from work, but if preindustrial medieval peasants enjoyed anywhere from eight weeks to an entire half year off,[277] we can throw a few extra holidays in. We will also enjoy regular events and performances from conventions and tourism. All holidays,

---

[277] Schor, Juliet B. The Overworked American: The Unexpected Decline of Leisure. New York, Basicbooks, 1993.

whether internationally recognized or unique to Augury, should be celebrated with enthusiasm and respect as a full spectacle. We can also benefit from local festivals dedicated to arts and artists; music and musicians.

Video Gaming & Creative Industries

With cheap electricity, fast internet and powerful computers (and limited outdoors to explore), Augury is a natural center for the global gaming industry (which is currently worth over $150 billion, even more than the music and movie industries combined[278]). With dedicated facilities like arcades and arenas, Augury could be perfectly poised to host gamers, tournaments, developers, studios, venues for performances of various kinds, and conventions like Comic Con, all of which bring in sizable income with minimal overhead cost. Tourists could enjoy world-class virtual-reality facilities, esports arenas which host regular tournaments, IMAX gaming theaters, a specially curated arcade of both new and games, and conventions like no other. And—what goes better with arcades than a bowling alley, mini golf course, laser tag arena, pizzeria, and movie theater (all styled like a classic 1980s era American mall, of course)? Augury should be an epicenter for these creative entertainment industries, which—with proper facilitation—will likely become Augury's economic backbone.

A government **business liaison agency** should seek out and encourage suitable businesses to establish operations in Augury. Suitable industries might include video game development, software development, entertainment (including film, music, writing, streaming, etc.), journalism, marine biology/technology, aeronautical engineering, and so on. Industries which will compete with executive departments should not be actively sought out; rather, qualified industry professionals should be recruited to improve those departments.

---

[278] Divers, Gavin. "Gaming Industry Dominates as the Highest-Grossing Entertainment Industry." Gamerhub, 24 Jan. 2023, gamerhub.co.uk/gaming-industry-dominates-as-the-highest-grossing-entertainment-industry/. Accessed 27 Apr. 2023.

Videogames are also powerful tools for scientific research and simulation.[279] This may lend itself to a unique research and development industry in Augury. We must never underestimate the power of play and fun in increasing our capacity to complete a task.

Entertainment Center

We acknowledge the importance of play for children (though sometimes we don't realize **how** important it is), but do we realize that it is also important for adults? "Playing is just as important for adults as it is for children. Among its many benefits, adult play can boost your creativity, sharpen your sense of humor, and help you cope better with stress[280]." Sometimes, our need for play has an outlet in the form of sports or video games—though without children around, we often neglect this need. In Augury, we must facilitate play for all ages. In the 1980s, indoor technology-based entertainment reached a peak, before home devices sent us back to our homes to enjoy our entertainment in isolation. I believe that it was better before, when you would go out to a bowling alley or an arcade, where you might see your friends or meet some new ones. Augury should have an "entertainment center" facility, outfitted with attractions like a bowling alley, mini golf course, arcade, laser tag arena, go-kart track, pizzeria, prize counter, food court, fountains you can wade in, esports arena, and a movie theater. These facilities should house both vintage and modern technology, but I think using the interior design characteristic of classic 1980's shopping malls, with neon and interesting shapes, would be thematically appropriate and visually appealing. People do love the "retro" aesthetic.

---

[279] Markoff, John. "In a Video Game, Tackling the Complexities of Protein Folding." The New York Times, 4 Aug. 2010, www.nytimes.com/2010/08/05/science/05protein.html. Accessed 19 July 2023.
[280] Vogel, Kaityin , and Marney A. White. "The Importance of Play for Adults: Tips for Being More Playful." Psych Central, 10 Nov. 2022, psychcentral.com/blog/the-importance-of-play-for-adults. Accessed 26 Aug. 2023.

Pop Culture Preservation Project

Augury is partially intended to function as a sort of "apocalypse bunker" refuge for catastrophic circumstances (which is why it must always be equipped to accept refugees), which is described in Part 3: Emergency Response, Apocalyptic Scenarios. This protective capacity can also be extended to human heritage. Since our very beginning, storytelling has been a defining trait of the human race. What started out as tales around a campfire has become a rich, infinite multi-genre cultural multiverse of literature, comic, movie, serie, and video game canons. The rich human culture must be preserved, but also appreciated. Therefore I propose we maintain in Augury a massive, exhaustive collection of pop culture, in any and all formats, so that we and our posterity can enjoy our small acts of creation. This collection should be on display and accessible to all citizens and visitors, possibly as an addition to the library or interspersed throughout the city. Means to view this media, of course, must also be preserved—digitizing is excellent and very convenient, but we should also have VCRs and record players and whatever else we need to use the originals.

Marine Biology & Oceanography

By catering to organizations like the NOAA, Universities, and other marine research, Augury can rent the best facilities on earth for marine biology. Airlocks will allow researchers to walk out onto the ocean floor to study reefs or plan expeditions into the deep. Holding tanks of various sizes can facilitate close study of specific organisms and biological relationships. Sustainable proximity alone is significantly advantageous towards the study of submarine environments. The Augury convention/exhibition center could become the venue of choice for marine biology and oceanography conventions.

Consider also collaboration with aerospace organizations like NASA—astronauts frequently train underwater.

Mining & Salvage

Extensive, unregulated undersea mining would be environmentally detrimental, but responsible mining with consideration given to long-term consequences will still allow for the allocation of some minerals, such as manganese nodules, salvage, and ocean water mineral extraction. "The bottom of the world's ocean contains vast supplies of precious metals and other resources, including gold, diamonds, and cobalt[281]." These deposits can often be found around undersea vents, which will be more accessible from Augury than ever before.

Economic System

There is a common sentiment circulating on the internet these days, centered around the tagline: "I do not dream of labor." Bootstrap capitalism has tricked us into believing "what did you want to be when you grow up" was a sinister question, that we should not aspire to work because the only work we know is to build the wealth of megalomaniacs. But labor is just another word for work, and work is nothing more than "effort performed to achieve a purpose." Labor is not good or evil; it is a tool which has been abused. We dream of a world where labor is not exploited at the expense of our survival. There is no reward in labor anymore, with the abuse from customers and bosses, long hours with insufficient compensation. When we and the people around us see the value of our labor, when we can enjoy its fruits, when it is for our good and the good of the people we care about rather than to inflate the wealth of another, labor is fulfilling and satisfying. Just look at what we did during quarantine—people took up new hobbies and projects they wouldn't have had time for otherwise. Despite being stuck inside we continued to create and produce.

---

[281] Howard, Brian Clark. "The Ocean Could Be the New Gold Rush." History, 13 July 2016, www.nationalgeographic.com/history/article/deep-sea-mining-five-facts. Accessed 31 July 2023.

At the end of the day, labor is, inherently, necessary for us to survive—whether we live in a city or off the land in a cabin in the middle of nowhere. Unless we are to live in isolation, we have to grow beyond the American "you don't owe anybody anything" mentality. Living in a community is collaborative, and we owe to each other what we are able to give. If we are to take part in a community, we must contribute to it. If we expect our neighbors to contribute to our community, we must not exploit their contributions, and we must value the crucial labor and laborers that make a community work. But that does not mean we must earn the right to survive—there will always be members of society who suffer from disability or some form of weakness which reduces or prevents entirely their ability to work, and if we claim to be a community, we must care for those who cannot care for themselves without any expectation of compensation.

The average wage gap ratio in America in 2022 was 670:1 (according to a study conducted by the Institute for Policy Studies), meaning the very top executives in large companies were compensated on average $670 for every $1 an entry-level worker received. Historically, this type of wealth inequality (and corresponding social status inequality) leads to instability and collapse—so in Augury, wealth inequality must be regulated to encourage a cooperative rather than competitive economy. Socioeconomic inequality should be minimized, and the wealth of citizens should fall on a bell curve to maintain balance. To maintain this balance, we will need to carefully design policy which protects it.

One of the most radical of these policies which I propose is the "No Employees Rule," which prohibits any private businesses from having employees—in other words, all workers in cooperation with a business organization must be a partner (receive a percentage of profits rather than a fixed rate and have some power of influence; like a co-op) or a freelance/contract worker (greater degree of freedom; not subject to company policies). By these standards, all companies must either be some variation of a cooperative, employee-owned business or a sole proprietorship.

The local private economy is intended to be kept small, never large enough to exert influence over any part of the government or policy in Augury. Small businesses are prioritized over larger franchises, and huge corporate chains are not allowed. Businesses should have a maximum size for the number of partners calculated based on the yearly census, so that no single business ever composes too large of a percentage of Augury's population. The same principle applies to total net income, to prevent a business from using financial income to form a monopoly, oligopoly, corporate oligarchy, or corporatocracy. When a business grows too large and approaches income or size ceilings, it will be required to either downsize or split-up. In some cases, if the industry is deemed crucial to the people of Augury and should therefore remain growing, the business may be nationalized so business practices and policies can be held publicly accountable . The intent of these policies are to prevent businesses or industries from exerting undue influence on citizens, and to protect small businesses.

Other economic policies include (but are not limited to) minimum wage requirements calculated yearly based on minimum cost of living, financial data transparency laws for all government departments, and wages for government employees, which are calculated yearly based on the minimum wage. If government employees (including department heads) want a higher wage, they have to raise the minimum wage.

Some may be concerned that this model for an economic system will reduce the reward of profit, without which citizens would have no motivation to be productive or excel. I would recommend considering the work of Wikipedia editors, open source coders, volunteer firefighters, Minecraft players, and many more. Human beings inherently want to make things, to grow, to improve—and will do so if given adequate opportunity.

"The desire to create is one of the deepest yearnings of the human soul."

—Dieter F. Uchtdorf

Nationalized Banking

Privatized banking has long been an industry based on greed and exploitation. Profits are reduced by fiscal responsibility, so it's in the interests of banks to enact policies like overdraft fees, which banks collected $7.7 billion of in 2022.[282] Instead, consider having banking run as a national service, like the postal or library service, where the motivation is the wellbeing of the individual rather than profit. Banking policies, offers, rates, and contracts can be publicly available and accountable, and changes be made by popular vote (moderated by economists to prevent economic crisis). The more transparent finances become, the less opportunity for exploitation. Integration of Minerva A.I. monitoring could also reduce theft, fraud, and laundering—with a court order, of course.

Augury's economy should use a centralized digital fiat currency[283] called **Aurichalcum**[284] (referred to as copper or coppers for short), the blockchain data for which will be stored on Apollo servers, but moderated and monitored by Minerva algorithms with as little human interaction as possible. Minerva systems will keep an overview of Aurichalcum units and their ownership and define whether new units can be created. If new units can be created, Minerva defines the circumstances of their origin and how to determine the ownership of these new units. If Minerva detects any issues, significant value fluctuations, or problematic trends, it will alert a human technician. Though Minerva tracks the movement of Aurichalcum movements, it is illegal for a human technician to track exchanges of currency between individuals without a Judicial warrant. Human economists

---

[282] Consumer Financial Protection Bureau. "Overdraft/NSF Revenue down Nearly 50% versus Pre-Pandemic Levels." Consumer Financial Protection Bureau, 24 May 2023, www.consumerfinance.gov/data-research/research-reports/. Accessed 6 July 2023.

[283] Rodeck, David. "Digital Currency: The Future of Your Money." Forbes Advisor, 31 Mar. 2021, www.forbes.com/advisor/investing/cryptocurrency/digital-currency/. Accessed 18 July 2023.

[284] "The Internet Classics Archive | Critias by Plato". classics.mit.edu. Paragraph 13. Retrieved 17 November 2021.

should monitor this system to ensure economic health as a redundancy, especially for the first decade or so until Minerva algorithms demonstrate effectiveness.

Citizens should be issued an ID card with a secure RFID chip which can be used as identification, a debit/credit card, passkey for certain doors, and redemption for online purchases or tickets, and certain perks afforded to citizens (for example, consider certain passageways unlockable only to citizens so they can avoid large crowds and heavy traffic from tourism). Uses of this card can be tracked only with a court order to protect privacy. The same credentials could also pair with a QR code loaded on their phone from their online government account. Of course, a fraud agency in Minerva to address identity theft would be necessary.

Tax Policies

Taxes in almost any society are a necessary evil, but in Augury, they will be designed to be as simple and low-maintenance as possible. Taxes should be deducted primarily (almost exclusively) from income, at the lowest rate possible to subsidize whatever costs the nationalized income can't cover.

Each year, citizens should receive an itemized tax report from the government, with the opportunity to contest any charges. There should never be any need to hire an accountant or worry about accidentally committing fraud on this report—if the numbers look right, all a citizen would need to do is file the report away for their records.

Tax code falls under the purview of policy determined using the MDDS— so if anyone doesn't like their taxes, they and everyone else can take a look at the numbers and crunch them a different way, and put their alternative to a vote. There will need to be a few hard policies in place to prevent economic collapse or inequality, but beyond that, why shouldn't we choose our own tax rate?

Politicians

The concept of a politician typically evokes the idea of a representative who votes and perhaps introduces new policies on your behalf. The Moderated Direct Democracy System eliminates the need for representative politicians—the closest thing in Augury will be Department Heads and the City Manager. Like the economy, politics in Augury should be shifted away from a competitive mentality and towards cooperation. These leaders must be gracious, humble, willing to learn from and work alongside fellow candidates. Their approach should be to work alongside other officials to solve problems, rather than to compete with each other for accolades and popularity. They should care more about the wellbeing of the city than their own personal ambition.

New Department Heads will be elected by citizens conditionally. Eligible candidates must have certain relevant credentials, experience, and education for any given role (as specified in each department description in Part 1). Consider having runner-up candidates become assistant Department Heads, to discourage bipartisanship. Political parties should be prohibited, as should donations from citizens or organizations towards political campaigns. Rather than making a spectacle of campaigning for office, positions of public office should be treated like the jobs they are. Positions in any branch of government can be held in contracts of 1 or 2 years at a time, with citizens voting to renew the contract at the end of the period or seek other candidates. Exact terms of the contract (as well as compensation and conditions for termination) should be publicly available. When candidates are being compared for a position, they should not campaign with rallies and parades—rather, they should make their qualifications publicly known, answer questions calmly and directly without manipulation, and if they take part in debates against their opponents, they should be conducted without a live audience. Candidacy for office in Augury should be a question of optimal qualification, not a performance for a popularity contest.

City Market

Imagine a large big-box supermarket like Walmart, Costo, or Sam's Club. In one stop, you can get groceries, furniture, tools, and visit an optometrist, pharmacist, and hair stylist. The experience is extremely convenient for the consumer, but can decimate the opportunities for small businesses in the area. What if we modified this model in a way that benefits both the customer and the vendor?

At first, Augury will only be populated by critical personnel and their families to keep the city running—but in time, we could see a community of local artisans and small businesses, creating and selling their goods. To facilitate that, I envision a facility not unlike a farmers market or an antiques mall, for a wide variety of goods and services from a variety of vendors. Augury will be unfriendly to large franchises, so small vendors can sell their goods without worrying about unfair competition, or the expense of their own brick and mortar store. A vendor could rent out a designated stall in a warehouse-like facility, next to other similar vendors (divided by category for the convenience of the customer). The vendor would take care of stocking and apply a special RFID barcode tag to their products, and would have the choice to stay with their shop or not. They can keep a shop open to customers, but spend their valuable time on something besides sitting in a stall all day. Market staff would take care of cleaning, checkout, and preventing shoplifting. Meanwhile, the customer has a one-stop-shop for all their needs. Hephaestus can maintain a section for locally manufactured goods; Demeter will stock the produce section. Any goods needed that can't/aren't produced locally can be imported and stocked on shelves, conveniently in the same facility.

To save citizens from inconveniently long walks, Augury should make extensive use of **vending machines** throughout the city. Modern vending machines are not just limited to candy, cheap snacks, and soda—there are machines which replace pharmacies, drugstores, and other small shops; some which will cook bread or a customized pizza in front of you, some which are refrigerated and hold fresh produce, cheeses, meats, dairy, eggs, or entire hot or cold meals,

canned food—the possibilities are endless. Vending machines make simple shopping easier and more efficient, in terms of space, time, and resources. Additionally, they remove the need for a human clerk to run checkout lines in a store. A well-built machine, reinforced with thick glass and bolted to the floor, would also reduce losses from theft. These machines could even be programmed to unlock in case of emergency, to serve as a repository of supplies. Partnering with Aegle, some vending machines could even offer supplies for rudimentary first aid. Medical personnel could access these supplies without payment.

Emergency Response

On the bottom of the ocean, emergencies need to be handled quickly and effectively. Emergency response teams in Augury should maintain and regularly drill contingencies and response plans for a variety of emergency scenarios, including but not limited to those listed below. Emergency response stations should be regularly interspersed throughout the city as well, containing supplies like antihistamines, first aid kits, defibrillators, and miniature scuba tanks.

Breaches

Augury will sit 30-40 meters or around 100-150 ft underwater. At this depth, the water pressure is between 4 and 5 atmospheres, which is 4 to 5 times the atmospheric pressure at sea level, or 59 to 73 pounds per square inch (which is around the same level of water pressure that should come out of a faucet). By maintaining a similar level of air pressure inside, Aether can reduce the danger of depressurization, but water will still get in if the interior pressure is at all lower and the exterior walls are breached. In this situation, the compartment would be evacuated, then sealed off at bulkheads. Next, the air pressure in the compartment will be increased above the ocean water pressure, so that air is pushing out rather than water pushing in. The breach can then be sealed with an industrial adhesive, and an electric current applied to the exterior wall to replace the mineralization. Once workers in Hestia are confident that

the breach is adequately patched, the compartment can be drained, dried, and reopened.

Fire

Fire in an airtight, hyperbaric environment is catastrophic, so the Augury fire department must work quickly to extinguish any fire before it spreads. Certain areas like electrical substation rooms should be outfitted with self-deploying fire extinguisher bombs as a contingency. Some smaller fires can be addressed by extinguisher turrets mounted in major rooms, concourses, and atriums, which use Minerva algorithms to automatically spray any hot spot that exceeds a certain temperature. Larger fires must be addressed in person, so firefighters will use a combination of chemical fire extinguishers and water pumped in from the ocean. If a fire grows truly out of control, the compartment will need to be evacuated and sealed at bulkheads, then deoxygenated. Most flammable fuels should be illegal or highly regulated; only importable by government departments like Hephaestus or Hestia.

Police

Augury's police force will be a relatively small organization of civil servants responsible for maintaining public order and safety. They rarely use force, and only ever as a last resort. For most public disturbances, police officers work with the department of Mentality, ready to provide support if necessary to social workers and therapists. Their jobs mainly consist of providing help to citizens and visitors in need, connecting them with Executive and Judicial departments as needed. They will not carry a firearm, as firearms of any kind should be universally illegal in Augury[285]. Instead, they should be trained in various forms of martial arts, and trained to de-escalate situations. If a situation demands it, they will use non lethal means to incapacitate aggressors—for example, as some

---

[285] See "Frequently Asked Questions."

Japanese police do, rolling them in large futons until they calm down[286].

Defense

Augury will not maintain a formal military and will not ever deploy a military presence on foreign soil. However, it should maintain remotely controlled defensive turrets mounted on exterior buildings in case of foreign attack, contingency plans in case of invasion, and a reserve National Guard trained to protect civilians and address emergency situations. These guard members would need dive training, to operate outside the city or if compartments of the city needed to be flooded for security purposes. Combat strategies should be a last resort.

Augury could also make use of remotely operated drones with powerful engines, designed to attach to an object and push it away from the city in case of collision course (whether from an antagonistic submarine, falling debris from above, etc.)

---

[286] Berteaux, Anthony. "How the Japanese Police Use Futons, Not Guns." Www.bbc.com, 1 June 2017, www.bbc.com/news/av/world-asia-pacific-38534288. Accessed 19 July 2023.

253

Apocalyptic Scenarios

The idea of an "apocalypse" of some kind may seem like a science fiction trope in popular media—in fact, the trope is so popular (and has been so popular since World War II and the Cold War) because we are so aware of the variety of threats to the continuation of the human race. Though most threats are unlikely to cause the extinction of human life, they are more than capable of greatly reducing our population or destroying vital infrastructure, reducing our access to vital resources. These natural and artificial "global catastrophic risks" include:

- Economic collapse[287]
- Electrical grid failure[288]
- World War[289]
- Thermonuclear holocaust (this has almost happened **accidentally** around 20 times since the invention of nuclear bombs[290])
- Radioactive fallout[291]

---

[287] Turchin, Peter. "America Is Headed toward Collapse." The Atlantic, 2 June 2023,
www.theatlantic.com/ideas/archive/2023/06/us-societal-trends-institutional-trust-economy/674260/. Accessed 30 July 2023.
[288] Ezrati, Milton. "America's Electric Grid Is Weakening." Forbes, 24 Mar. 2023,
www.forbes.com/sites/miltonezrati/2023/03/24/americas-electric-grid-is-weakening/. Accessed 30 July 2023.
[289] Rosenfeld, Gavriel . "Perspective | How Will We Know When It's World War III?" Washington Post, 5 Apr. 2022,
www.washingtonpost.com/outlook/2022/04/05/how-will-we-know-when-its-world-war-iii/. Accessed 30 July 2023.
[290] Wikipedia Contributors. "List of Nuclear Close Calls." Wikipedia, Wikimedia Foundation, 20 Dec. 2019, en.wikipedia.org/wiki/List_of_nuclear_close_calls. Accessed 27 Aug. 2023.
[291] US EPA,OAR. "Radioactive Fallout from Nuclear Weapons Testing | US EPA." US EPA, 30 Nov. 2018,
www.epa.gov/radtown/radioactive-fallout-nuclear-weapons-testing. Accessed 4 Sept. 2023.

- Asteroid/comet impact event[292](like the Bennu asteroid, which has the highest cumulative rating on the Palermo Technical Impact Hazard Scale and has a small chance of impacting earth in the next 200 years[293])
- Supervolcanic eruption[294]
- Natural pandemic from antibiotic-resistant "superbugs[295]"
- Artificial pandemic (intentional or accidental; biological warfare/bioterrorism, which is becoming more accessible[296])
- Coronal mass ejections like the 1859 Carrington Event[297] (electromagnetic/gamma solar emissions)
- Environmental change (rising global temperatures leading to famine, drought, wildfires, flooding, etc.)[298]
- Flooding of coastal cities from rising sea levels[299]

In addition to the hypothetical, we have record of countless "population crash" events in history which decimated the human

[292] Brain, Marshall, and Sarah Gleim. "What If an Asteroid Hit the Earth?" HowStuffWorks, 4 Dec. 2007, science.howstuffworks.com/nature/natural-disasters/asteroid-hits-earth.htm. Accessed 30 July 2023.
[293] Wikipedia Contributors. "101955 Bennu." Wikipedia, 13 Oct. 2023, en.wikipedia.org/wiki/101955_Bennut. Accessed 16 Oct. 2023.
[294] Sparks, Stephen. "Supervolcanic Eruption – Overview (Extract Global Catastrophic Risk Report 2022)." Global Challenges Foundation, 2022, globalchallenges.org/library/supervolcanic-eruption-overview-extract-from-glo bal-catastrophic-risk-report-2022-2/. Accessed 30 July 2023.
[295] Papa, Sophia. "Antibiotic-Resistant Superbugs: A Looming Health Threat." Www.ucihealth.org, 3 Mar. 2023, www.ucihealth.org/blog/2023/03/antimicrobial-resistant-superbugs. Accessed 26 Nov. 2023.
[296] Esvelt, Kevin M. , and Cassidy Nelson. "Sources —Biorisk." Kurzgesagt—in a Nutshell, 2 July 2023, sites.google.com/view/sources-biorisk. Accessed 27 Aug. 2023.
[297] Homeier, Nicole, and Lisa Wei. Solar Storm Risk to the North American Electric Grid. 2013.
[298] NASA. "The Effects of Climate Change." Global Climate Change: Vital Signs of the Planet, NASA, 2023, climate.nasa.gov/effects/. Accessed 27 Aug. 2023.
[299] NOAA. "2022 Sea Level Rise Technical Report." Oceanservice.noaa.gov, 2022, oceanservice.noaa.gov/hazards/sealevelrise/sealevelrise-tech-report.html. Accessed 27 Aug. 2023.

population, including the following examples (with some of the highest death tolls):

From man-made conflict:

- World War II (70-118 million, roughly 73% civilian[300])
- Reign of Mao Zedong (~66 million, purges, mostly famine[301])
- Genghis Khan's conquests (~40 million[302])
- British India (~27 million, mostly famine due to European business policies[303])
- Atlantic Slave Trade (~16 million[304])
- Mideast Arab Slave Trade (~18.5 million[305])
- Conquest of Native Americans (~15 million from disease, famine, and genocide[306])
- Fall of Rome (~7 million[307])

From pandemics[308]:

- Black Death (75-200 million, 1346–1353)
- Spanish Flu (17-100 million, 1918–1920)

---

[300] "Defense Casualty Analysis System—World War II." Dcas.dmdc.osd.mil, USA.gov, dcas.dmdc.osd.mil/dcas/app/conflictCasualties/ww2.
[301] Courtois, Stéphane, and Mark Kramer. The Black Book of Communism : Crimes, Terror, Repression. Cambridge, Mass. ; London, England, Harvard University Press, 1999.
[302] Macfarlane, Alan. The Savage Wars of Peace: England, Japan and the Malthusian Trap. Springer, 19 Nov. 2002, p. 50.
[303] Davis, Mike. Late Victorian Holocausts. Verso Books, 2017.
[304] Stannard, David E. American Holocaust : The Conquest of the New World. New York, Oxford University Press, 1993.
[305] Segal, Ronald. Islam's Black Slaves : The Other Black Diaspora. New York, Farrar, Straus And Giroux, 2002.
[306] Stannard, David E. American Holocaust : The Conquest of the New World. New York, Oxford University Press, 1993.
[307] Gibbon, Edward. The History of the Decline and Fall of the Roman Empire. Vol. 1-3, Palala Press, 8 May 2016.
[308] Wikipedia Contributors. "List of Anthropogenic Disasters by Death Toll." Wikipedia, 29 July 2023, en.wikipedia.org/wiki/List_of_anthropogenic_disasters_by_death_toll#Diseas e_and_famine. Accessed 30 July 2023.

- HIV/AIDS Crisis (40.1 million, 1981–present)

From natural disasters[309]:

- 1931 China Floods (~4 million, 1931)
- 1887 Yellow River flood (~2 million, 1887)
- 1976 Tangshan earthquake (~655K, 1976)

From famine[310]:

- Great Chinese Famine (11-40 million, 1959–1961)
- Chinese famine of 1906–1907 (25 million)
- Northern Chinese Famine of 1876–1879 (9-13 million)

These are just a few examples of potential threats to humanity's existence, and enough to make many of us deeply paranoid of an impending disaster of some kind[311]. The unfortunate truth is that an apocalypse of some kind is not only possible, but growing increasingly probable. The world is unpredictable, and disaster is a theme of history, not fiction—and exacerbated by available technology.

Fortunately, there is a bright side. Once fully constructed, Augury will be the perfect "doomsday shelter." Underground doomsday shelters and bunkers have been a popular construction project for people who call themselves "preppers" around the world who anticipate that environmental changes or impending nuclear war will soon force them to live underground. The video game franchise "Fall-Out" centers around this concept. Naturally, these subterranean complexes are prohibitively expensive, and therefore are only accessible to the wealthy—they are typically rented by the larger supplier companies that build them. The founder of one such company, the Survival Condo Project, noted that "When the rhetoric

---

[309] See previous footnote.
[310] See previous footnote.
[311] Cole, Rachel. "10 Failed Doomsday Predictions". Encyclopedia Britannica, 3 Apr. 2013, https://www.britannica.com/list/10-failed-doomsday-predictions. Accessed 31 July 2023.

was hot and heavy between Trump and Kim Jong-un in North Korea, there was a real big spike in phone calls[312]."

According to Business Insider[313], an effective shelter needs the following:

- Stockpiles of supplies like imperishable food and medical supplies
- Protection from flooding
- Distance from military targets
- Thick walls to keep out radiation, contaminated water and air, and protect from explosives
- Airtight walls or filtration systems to keep out radiation
- Security systems/measures
- Sustainable agricultural arrays
- Access to potable, uncontaminated water
- Localized electrical grid/power source

In short, an effective doomsday shelter needs to be a self-sufficient off-grid homestead completely sealed off from the rest of the world. Augury will fulfill all of these needs and more, and be far more comfortable than any subterranean bunker. As many works of fiction have explored, aquatic refuge is optimal for surviving global catastrophe[314]. Survival shelters are built underground to protect them from heat, shockwaves, and radiation behind a thick layer of earth. Water is also an effective barrier against these threats. Water is a potent insulator to radiation and heat; the thermal inertia of the ocean keeps most of it much cooler than the surrounding atmosphere even in the event of a period of natural or artificial global

---

[312] Larry Hall to Business Insider.
[313] Bendix, Aria. "Survivalists Are Buying Underground Doomsday Bunkers to Prep for the Apocalypse. Here's What They Look Like." Business Insider, 18 Sept. 2019,
www.businessinsider.com/what-makes-a-good-doomsday-shelter-2019-9. Accessed 30 July 2023.
[314] Turchin, Alexey, and Brian Patrick Green. "Aquatic Refuges for Surviving a Global Catastrophe." Futures, vol. 89, no. 0016-3287, May 2017, pp. 26–37, www.sciencedirect.com/science/article/pii/S0016328716303494, https://doi.org/10.1016/j.futures.2017.03.010. Accessed 18 May 2021.

warming. Though ocean water at 100-150 feet deep is not cold enough to make Augury unlivable, the aforementioned nuclear decay heat (or electrical heat) can be used to make Augury comfortable. Only 13.8 feet of water is needed to reduce gamma radiation intensity by a factor of a billion[315]—at 100-150 feet, we will still be able to farm, fish, and dive near the ocean floor without risking exposure. Water is also very effective at suppressing the effects of explosive shockwaves[316]. With the exception of a direct strike, Augury would be immune to nuclear holocaust. Water also allows for a shelter to be built deeper, farther from the surface—the ocean is cooler at depth, whereas the earth gets warmer, eventually requiring cooling systems. This will protect Augury from even extended exposure to nuclear or cosmic radiation.

Isolation is also easy to achieve within Augury—with geographic isolation and limited entry-points, the city can be easily sealed off from the surface. By design, Augury must be self-sustaining, so it will be able to function normally without air, water, food, medicine, or power from an outside source. Water and air will be internally purified and recycled, preventing the risk of contamination. Of course, these vital supplies should be stockpiled in case of equipment failure, unexpected system loads, or emergency. As previously mentioned, Augury should be built with extra capacity to accept refugees; further additional supplies will be necessary for this case. In the event of a global pandemic, quarantine can be easily enforced.

Many futurists such as Elon Musk have (urgently) suggested emergency extraterrestrial egress to the Moon and Mars[317], but an aquatic refuge would be far more sustainable and indescribably more

---

[315] Amir, Nofit. "Effective CBRN Radiation Shield." StemRad, 31 Oct. 2018, stemrad.com/cbrn-gamma-radiation/.
[316] Salter, Stephen, and John Parkes. "Why Is Water so Efficient at Suppressing the Effects of Explosions?" The Journal of Conventional Weapons Destruction, vol. 22, no. 1, 7 May 2018, commons.lib.jmu.edu/cisr-journal/vol22/iss1/9/. Accessed 30 July 2023.
[317] Mosher, Dave , and Kelly Dickerson. "Elon Musk: We Need to Leave Earth as Soon as Possible." Business Insider, 10 Oct. 2015, www.businessinsider.com/elon-musk-mars-colonies-human-survival-2015-10. Accessed 30 July 2023.

economical. A colony like Augury could safely function for generations in near-total isolation while still using earth's natural resources. Not to mention, the proximity would allow us to monitor as surface conditions improve.

There are several phenomena that consistently contribute to societal collapse that have caused empires to fall in the past which we must be sure to prevent and protect against to maintain Augury's stability:

- War/foreign conquest
- Pandemic/Famine
- Discontent among citizens
- Political/ideological division
- Overextension/Decentralization of power/weak leadership
- Economic Collapse/Economic inequality
- Poor adaptability/slow reformation
- Cultural assimilation/overwhelming
- Class inequality/Government catering to upper class rather than all citizens

Frequently Asked Questions

What is Augury? And what is its purpose?

Augury is a mass-capacity, self-sustaining submarine habitat. The purpose of Augury is to facilitate a community that prioritizes the needs for the physical, mental, and spiritual health of its citizens so that they will pursue self-optimization and self-actualization.

Where will Augury be built?

Augury couldn't be built just anywhere. To function as intended, the Augury building site needs two basic things: resources and sovereignty. Access to resources is flexible, as more plentiful resources can sometimes compensate for scarcer resources. But things like clean water, an energy source, and the ability to grow food are crucial. If energy is plentiful enough, it can be used to power grow lights, so access to sunlight is not an absolute requirement. Resources that can be sold for profit would also be highly beneficial, but what's most important is that the space is not already claimed by another governing body, so that Augury can be self-governing, self-regulating, and free to pursue the goal of prioritizing health by any means deemed appropriate, without being subject to foreign governments or culture. This limits available space to "Terra nullius," meaning land on earth not currently claimed, of which most options are unclaimed because they are unlivable. Though this seems discouraging, there is one environment that fulfills and exceeds the requirements, in the most plentiful natural resource on the surface of the earth: the ocean.

Augury will be situated off the eastern coast of America or near Bermuda, in waters shallow enough to support significant marine life, and where the ocean pressure, even on the floor, is not extreme. This location is in the middle of the Gulf Stream which is useful for power generation; it's close to transatlantic telecommunications

cables; it's close to shipping lanes between America and Europe which can facilitate trade; proximity to Bermuda which facilitates significant tourism; the depth allows sunlight to reach down, the location in general allows for propagation of marine plants like kelp (which provide both edible and commercial resources); and the ocean offers several commercial resources like manganese nodules and salvage.

This may come as a surprise, but I often think to myself that I would almost prefer to build Augury on land. Surely it would be simpler and easier, to live as humanity has always lived, in an environment we are biologically suited for, without need for specialized equipment and technology just to survive. The point is moot, however, because there is no land left unclaimed on the surface of the earth. There is no land we could claim with true sovereignty, at least not without paying dearly. The ocean is the last frontier on Earth. And after all, as I have explored throughout this manuscript, living in the ocean comes with advantages we wouldn't dream of on land. I predict that while it will be expensive and difficult to set up Augury on the front end, in the long run, it will be a better and more efficient way of life for modern humanity.

How will you build a city underwater?

There are two big obstacles to underwater construction which have so far prevented it from becoming popular: pressure and inaccessibility. Inaccessibility is obvious: traditional construction relies on infrastructure like roads and existing utilities. Underwater construction requires specialized equipment and workers. Barges must transport those workers and equipment as well as construction materials, which are exponentially more expensive when they are difficult to transport. Then add diving equipment, the cost of underwater welders, specialized cranes, and it's clear why underwater buildings aren't more popular. This is not even to mention that the water requires that buildings be constructed many times thicker and stronger to withstand the pressure, further compounding the cost.

There is, however, an alternative in the form of technology. In 1976, an architect named Wolf Hilbertz discovered that by passing electric currents through saltwater, over time a thick layer of various minerals similar to limestone will accumulate on the cathode. He patented this process under the name "Biorock". This process of mineral aggregation can form walls of stone thick enough to withstand ocean pressure around a frame of any electrically conductive material like rebar, using nothing but an electric current, which can be generated by the motion of the ocean itself. In addition to being inexpensive, passive, and mechanically simple, Biorock also cleans ocean water from dissolved minerals, adds hydrogen and oxygen to the water which encourages marine life, repels sharks due to its electric field, and is self-healing as long as the electric current remains active. Furthermore, Biorock is currently used to restore coral reefs because the surface of the aggregated minerals is perfect for coral growth. The entire exterior surface of Augury would eventually grow into a coral reef, further encouraging the health of the surrounding ocean, preserving marine life by serving as an artificial marine nature preserve, facilitating tourism and marine biology.

How long will it take to build?

Biorock mineralization rate depends on the amount of electrical power supplied. More power means faster but weaker mineralization; less power makes a slower but stronger shell. Based on past applications of biorock, I estimate 3-5 years to grow the exterior structure.

Won't it run out of air?

The Aether department's most crucial function is to balance the concentrations of oxygen and carbon dioxide by reoxygenating constantly to compensate for human respiration. Access to sufficient oxygen improves cleansing and tissue repair in the body, and helps them exchange gasses more easily. Insufficient oxygen supply can

result in serious health problems, but too much oxygen can cause issues as well.

Reoxygenation is primarily achieved by algae and phytoplankton tanks like the ones shown to the left, in which case Aether works closely with Demeter. Through its natural photosynthesis, marine algae can remove toxic gasses like carbon monoxide, nitrogen oxides, and sulfur oxides, with a success rate of over 98%. The algae metabolize these gasses and infuse oxygen into the filtered air, even more effectively than trees.

In case all these systems fail, air can also be pulled in from the Primary Induction Lighthouse.

How will the government work?

Augury's government is divided into the same three major branches as the American government: Judicial, Executive, and Legislative. The functions of these three branches, however, are significantly different.

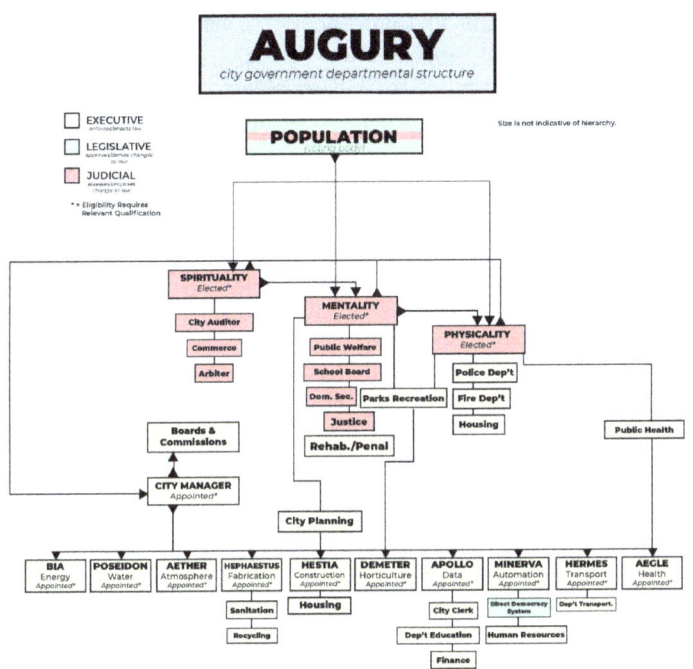

AUGURY
city government departmental structure

The Executive branch, comprised mainly of the ten vital utility departments, is responsible for doing the work necessary to keep the city running properly, anything from cleaning the air to expanding the city to parks and recreation.

The Judicial branch, divided into the three departments of Spirituality, Mentality, and Physicality, functions as a sort of auditing system. Their responsibility is to inspect and analyze the Executive branch and the city at large to ensure that the city-wide directive of prioritizing health is optimized. They constantly search for flaws or opportunities to improve and propose solutions to keep Augury at its best.

The Legislative branch, which comprises the entire population of Augury, is responsible for agreeing or disagreeing to any policy changes the Judicial branch proposes. Using the MDDS, Minerva may adjust the weight of an individual vote higher or lower to compensate for factors like expertise or conflict of interest.

How will Augury affect the natural environment?

Environmental preservation is extremely important to the spirit of Augury and the foundational principle of sustainability. To that end, the entire structure itself will greatly benefit the ocean around it. Biorock, the primary building component of Augury, cleans ocean water from dissolved minerals, adds hydrogen and oxygen to the water which encourages marine life, and repels sharks due to its electric field. Furthermore, Biorock is currently used to restore coral reefs because the surface of the aggregated minerals is perfect for coral growth. Therefore, the entire surface of Augury will eventually grow into a coral reef, further encouraging the health of the surrounding ocean, preserving marine life by serving as an artificial marine nature preserve, facilitating tourism, marine biology, and marine conservation awareness.

In addition, the closed system of Augury will force citizens to be conscientious of pollution and environmental conditions, which will facilitate relevant research and development towards reducing, reusing, and recycling waste. Necessity breeds invention!

Wouldn't everyone get claustrophobic?

In 2021, Americans spent 90% of their time indoors, and around 85% lived in densely populated urban areas. Though the effect on mental health from this lifestyle is concerning, it's clear that at least most American people are well accustomed to life indoors. This being said, Augury will include several expansive atriums to combat claustrophobia and offer citizens a wide-open space to unwind and relax. Demeter will maintain a high density of plant life in open areas

both for mental health and air quality. Citizens will be free to exit the city and return to the surface whenever they please.

How much will it cost to build Augury?

It's hard to accurately estimate the cost of a project so large when no other directly comparable project has ever been attempted. But taking into account the costs of off-shore drilling platforms and similarly-sized buildings on land, we estimate it cost around $150-$200 million (USD) to build the prefatory structure—after initial funding, nationalized income like underwater data centers will pay for ongoing expansion of the city and other expenditures.

Will guns be allowed?

Firearms or explosives of any kind will be universally prohibited in Augury. In a pressurized underwater container, a stray bullet could cause significant damage. While a bullet fired from a handgun wouldn't be powerful enough to penetrate an exterior wall, it could weaken a window, rupture a pipe, damage wiring or machinery, or might cause a deafening echo. Additionally, most interior walls will not need to be hard or thick enough to be bulletproof, so a stray bullet could travel much farther than intended.

Won't it be like Bioshock™?

The fictional underwater city of Rapture, which serves as the setting for the 2007 award-winning game series "Bioshock" developed by 2K and Irrational Games, collapsed into a localized apocalypse for many reasons—none of which depended on it being underwater. The first and most important factor was authoritarian fascism. Rapture's government was overbearing and totalitarian, overregulating some freedoms and under-regulating others, allowing the most powerful members of society to prey on the weakest with little to no accountability—eventually resulting in a stark wealth disparity. The

second was isolation—both physically, as Rapture was miles underwater, and politically, as citizens were not allowed to leave the city. Finally, Rapture was overrun by an addictive drug market. Rampant addiction will cripple any community—particularly if that addiction gives you uncontrollable aggression and superpowers through genetic mutation (and is coupled with a self-defense arms race to balance the playing field). Combine those main issues with a lack of cultural morality, demonization of empathy, a propaganda-induced general cult mentality, and an unregulated, free-market, purely individualist capitalist economy, and you can see clearly that Rapture was doomed from the start. If you're interested in the whole story, I highly recommend the 2011 prequel novelization "Rapture," written by John Shirley. Suffice to say, the similarities between Rapture and Augury are limited to being underwater, and taking names from the Greek Pantheon. The story of Bioshock is a critique[318] of Randian Objectivism[319]; the city being underwater was merely facilitative to that point. Rand was right in identifying some of the big problems, however:

> *"When you see that in order to produce, you need to obtain permission from men who produce nothing—When you see that money is flowing to those who deal, not in goods, but in favors—When you see that men get richer by graft and by pull than by work, and your laws don't protect you against them, but protect them against you—When you see corruption being rewarded and honesty becoming a self-sacrifice—You may know that your society is doomed."*
>
> —Ayn Rand, Atlas Shrugged

---

[318] "BioShock's Critique of Ayn Rand & Objectivism." Giant Bomb, 18 Apr. 2011, www.giantbomb.com/profile/ubik/blog/bioshocks-critique-of-ayn-rand-objectivism/80374/. Accessed 11 Aug. 2023.
[319] "Overview | AynRand.org." AynRand.org, 2019, aynrand.org/ideas/overview/. Accessed 11 Aug. 2023.

Doesn't this plan seem a little overambitious?

Sure, it does. Augury is a big idea. If implemented according to this plan, it could change the course of history. It will absolutely not be easy to do. It's an enormous construction project—but we undertake enormous construction projects around the world all the time. Consider the millions that go into building cruise ships, offshore oil platforms, and skyscrapers—of which there are more every year. The question isn't really whether we're capable of building something like this—we are. The big challenge will be convincing people it's worth the effort. Human beings are afraid of and resistant to change and the unknown more than anything else. We have an overwhelming tendency to stick with familiar suffering rather than risk a potentially better alternative if it's unfamiliar. I can write tens of thousands of words to convince people that Augury is the best cumulation of solutions to the big problems we face in the world, but the one thing I can't convince anyone of is to care if they don't. But the ocean is a frontier, and though we are characterized by our fear of change, we are characterized even more so by our curiosity. What is humanity if not a race of pioneers and explorers?

Eventually, some futurist company or entrepreneur is going to colonize the ocean. With modern technology it's only a matter of time before someone decides there's enough of a market for it and builds an ocean floor vacation destination. Should we wait for them to do it first and pay $499.99 a night to visit? Or should we do it ourselves and do it properly, and build the kind of community we're all desperate for?

Here is my personal perspective: like all Christians, I believe in providence. Augury strikes me as an idea much bigger than myself, considering my lack of qualifications—but does God not often work through humble means to accomplish great things and demonstrate His power, like using Gideon to free Israel from the oppression of the Midianites? I do not know if Augury is God's plan for me or just my delusion of grandeur or method of escapism. The way I see it, if Augury is not God's plan for me, then it may not be built after all, but

269

I will have learned from and enjoyed the process of planning it immensely. On the other hand, if Augury *is* God's plan for me, for us, then nothing could possibly stop it.

Opportunities for Expansion

"What goes too long unchanged destroys itself. The forest is forever because it dies and dies and so lives."

—*Ursula Le Guin; Tales from Earthsea: Dragonfly*

As Augury expands, it's important to plan for long-term sustainability with longevity and future growth in mind. Good civil engineering and planning is paramount. Just as Joseph William Bazalgette insisted on doubling the proposed size of London's sewers in the 1850s to accommodate the growth of the city[320], Augury should be built to handle a higher capacity than anticipated. New facilities can follow the same plan as outlined in Part 2, or whatever refined version is eventually used as standard building protocol. Structures should extend outwards along the paths of mass transit like spokes from a wheel, leaving room for ocean life and views in between. Facilities should never be constructed too densely; all living quarters must have windows into the ocean or into an atrium. Facilities of all kinds should be wrapped around central atriums. A section of Augury should always be empty (stop at around Step VIII) for storage and to accommodate emergency refugee camps. Not all new facilities need to be built as expansions onto the main structure; separate buildings will reduce the risk of cascading system failure and can be connected by "bathyducts" (like an underwater skyway). Some "suburban" communities in smaller compounds can be constructed in the outskirts of the city to have a more direct access to agricultural

---

[320] Collinson, Alwyn . "How Bazalgette Built London's First Super Sewer." Museum of London, 26 Mar. 2019, www.museumoflondon.org.uk/discover/how-bazalgette-built-londons-first-super-sewer. Accessed 31 July 2023.

interests like kelp forests and seagrass fields; this may appeal more to some people who prefer a more small-town or village dynamic.

Augury's community should not be allowed to grow too large; human beings have demonstrable limits of social capacity, and significant research points towards a concept similar to Dunbar's number, which "is a suggested cognitive limit to the number of people with whom one can maintain stable social relationships (around 150; anywhere from 100-200 depending on the person)—relationships in which an individual knows who each person is and how each person relates to every other person[321]." The larger Augury or any given community grows, the less functional its social health will be.

If the population grows beyond 5,000 or so, a "sister city" of sorts should be established nearby—however, Augury's population should always grow slowly, to ensure sustainability. The strongest trees grow the slowest. Smaller communities are healthier, as anonymity is reduced and people can more easily form relationships with their neighbors. Additionally, smaller communities are friendlier to small businesses, less expensive, and have lower crime rates[322]. Higher population density strains our capacity for empathy and our capacity to effectively and fairly govern.

If Augury's systems are well-designed enough and produce a significant surplus of vital resources, they can be exported around the world as humanitarian aid. If the concept of Augury proves itself in time to be fully viable and fulfills my expectations, more underwater or floating colonies like it may be built. The ocean has far more livable space than the surface[323], and if enough of the human

---

[321] Wikipedia Contributors. "Dunbar's Number." Wikipedia, Wikimedia Foundation, 31 July 2020, en.wikipedia.org/wiki/Dunbar%27s_number. Accessed 1 Sept. 2023.
[322] Callaghan, Julie Keller. "The Benefits of Living in a Small Town - Hutchinson Consulting." Hutchinson Consulting, 26 Aug. 2022, hutchinsonconsulting.com/the-benefits-of-living-in-a-small-town/. Accessed 31 July 2023.
[323] National Oceanic Atmospheric Administration. "How Much of the Ocean Has Been Explored? : Ocean Exploration Facts: NOAA Office of Ocean

population eventually relocates to the ocean, the surface environment may be able to recover from overfarming, pollution, and deforestation to pre-industrialization conditions.

Even if the monumental project of building Augury is completed within our lifetime, it will not be perfect. It will never be perfect. Augury will not be fair, or balanced, or efficient right away. It will take constant, ongoing work. Our systems, mechanical and social, will always be subject to improvement. Our people and culture must always be open to change and growth for the greater good. We must never, even as generations pass, get stuck in ruts of tradition and habit. Perfection is a horizon to reach for, not a prize to hold in our hand[324].

> "Westerners are fond of the saying 'Life isn't fair.' Then, they end in snide triumphant: 'So get used to it!' What a cruel, sadistic notion to revel in! What a terrible, patriarchal response to a child's budding sense of ethics. Announce to an Iroquois, 'Life isn't fair,' and her response will be: 'Then make it fair!'"
>
> — *Barbara Alice Mann*

—

> *"What is it that the child has to teach?*
>
> *The child naively believes that everything should be fair and everyone should be honest, that only good should prevail, that everybody should have what they want and there should be no pain or sadness.*
>
> *The child believes the world should be perfect and is outraged to discover it is not.*
>
> *And the child is right."*
>
> — *Tzvi Freeman, Wisdom to Heal the Earth - Meditations and Teachings of the Lubavitcher Rebbe*

Exploration and Research." Oceanexplorer.noaa.gov, oceanexplorer.noaa.gov/facts/explored.html. Accessed 31 July 2023.
[324] Adapted from a quote by Niel deGrasse Tyson while voicing Waddles the pig in Gravity Falls, trademark of Disney Enterprises, Inc.

An Average Day

Let's imagine a possible future. Imagine Augury is funded and built, established as a colony and a community. Let's say you decide to move there. You apply, pass citizenship requirements, and settle into the city that will be your new home. What can you expect from your new life underwater? What would an average day living in Augury look like? Well, depending on your personal lifestyle, it might look something like this:

You wake up to a spectacular ocean view. The windows in your apartment give you a stunning view of the ocean floor through clean (but not sterile) water. The outer surface of the city has coral growing on it, so you are likely greeted every morning by some curious fish or other sea life. Throughout the city, you may see jellyfish drifting past a window or hear distant whalesong. You can see the surface above you, giving you an idea of the weather up there. Of course, down here, the interior atmospheric conditions are completely predictable and stable thanks to Aether.

Since the city systems have so many built-in automations, redundancies, cushions, and safety nets, things run on a pretty leisurely schedule, so you probably wake up when you feel like it. Maybe you wake up early and go for a jog before breakfast. Maybe you're a night owl so you sleep in. Augury functions more around deadlines for tasks than hours in shifts, so "regular hours" are a relative concept. You decide the schedule you will operate on.

Your home is arranged to your liking—including the layout of the rooms. Your apartment is comfortably sized, laid out to your preference—not small or compact, but modest enough to use space efficiently—inspired by Japanese space-efficient interior design (but not as cramped). Cabinetry spans floor-to-ceiling to make the most effective use of limited space, some furniture is modular or multipurpose, some furniture folds, and appliances are multi-function. You don't have an excess of wealth or luxury and your apartment is not built for the runaway accumulation of possessions. But the possessions you have are well-made, you have lots of shelf room for books or movies or keepsakes, a comfortable living room with a very nicely-sized screen, and maybe even a balcony overlooking one of the atriums. The walls are thick enough to block out all the sound from your neighbors and you have complete control of the temperature. Your home is cozy and has everything you need.

You rent your apartment, but so does everyone else (including even the highest government officials)—privately owned real estate can quickly get out of hand in a closed system like Augury, especially when every structure is built and maintained by a public government department. But despite the apartment not being your personal property, you have strong rights to the space you live in. You can customize it to your heart's content with furniture, paint, and fixtures, and even make some modifications to the layout with modular prefabricated panels. Repairs are prompt and you never have to worry about unannounced visits from your landlord.

Next, a healthy breakfast. Your kitchen is efficient and fully equipped for you to cook whatever you might like for breakfast. Fruits, vegetables, even poultry and eggs, as well other fresh foods that can be grown locally are available from Demeter at their perpetual farmer's market, which is within walking distance if you forgot to stock your fridge. Other, more exotic ingredients might be available at the main market. Non-perishables can be ordered online and imported.

Or, perhaps, if you don't feel like cooking, you might decide to visit a local restaurant or bakery for breakfast. You sit at a small table off to

the side of an atrium, sipping a hot drink and listening to the birds and a babbling brook while you wait for your food.

After breakfast, it's time to get some work done. Work in Augury more often than not looks like a list of tasks that must be completed that day, and after you complete them, the rest of the day is yours to spend as you choose. If you work in one of the government departments, your daily work will typically take around 4-6 hours. Some days it will be more, but some days it might be even less. You work only when work needs doing.

Perhaps you're good with people, and the department of Mentality employs you to provide therapy or counseling to your fellow citizens—or Aegle as a nurse or doctor. If you're good with machines, maybe you just have some routine maintenance or cleaning of machinery with Bia, Hephaestus, Poseidon, or Aether, which is crucial to keep the city running smoothly. Have a green thumb? Maybe you're planting fruit trees with Demeter. Have a mind for the bigger picture? You may just be in charge of taking a look at Augury at large with Physicality, Mentality, or Spirituality, and coming up with ideas to help the city run better. Or maybe you're performing one of the most important jobs in any community by allocating and recycling garbage with Hephaestus. Of course, if you decide your current job isn't for you, you can switch to a different one—as availability allows. Perhaps you choose the private sector route and offer goods or services in a business. Augury is friendly to small businesses and tries to help local shops and restaurants thrive. From teaching to building, programming to medicine, whatever you do, your work is valued and your community appreciates you—because your work builds a community where people can thrive and be healthy. "Far and away, the best prize life has to offer is the chance to work hard at work **worth doing**."[325]

After just a few hours of work, it's time for you to do what recharges your battery, what brings you joy, or what makes you feel alive. Most people in Augury are pursuing some kind of personal project in their

---

[325] Theodore Roosevelt, 1903

spare time—what's yours? Is it creating art? You might just find a blank wall or floor or even ceiling somewhere in the city that needs a mural painted on it and be commissioned by the government. Are you a music lover? Share your passion on public pianos placed throughout the city, or from a park bench. Do you prefer to spend your time consuming art made by others? With one of the largest libraries of published and unpublished works in the world, you could spend several lifetimes in the library (open late), arcade, and movie theater—reading, watching, or playing through worlds others have created.

Perhaps you prefer the practical arts—Hephaestus maintains open facilities where citizens can use specialized tools for woodworking, metalworking, and more. If you're a programmer or developer, Minerva and Apollo facilities can help you build entire virtual worlds, and they offer special incentives to game development studios and software developers who call Augury home. Do you love to work in charity? Spirituality can help with that, and connect you with people and causes that need help. Are you the athletic type? Check out our gyms, running tracks, racquetball and tennis courts, martial arts classes, and more—as well as other activities organized by the Parks and Recreation department. They maintain facilities for specific kinds of recreation as well as organize city-wide community events like galas, concerts, festivals, holidays, and more. They're also whom you should talk to if there's an activity you need help facilitating or want to request.

Do you want simply to spend time with friends and family? In Augury, areas like parks and libraries, movie theaters and arcades, gardens and markets, and even simple hallways have comfortable lounge areas for you to simply exist in with no expectation of payment. Gather your picnic, book club, sewing club, or family reunion in public spaces (which are, by and large, open 24/7). The city belongs to its people! A healthy community requires the capacity to comfortably coexist. Just stay safe and keep an eye on those maximum occupancy requirements and noise ordinances or you may be asked to relocate—being part of a community also means being considerate of others' safety and comfort.

Don't forget your chores! Augury may seem utopic, but it's just a normal community of normal people who still have to take out the trash and do their laundry. Fortunately, with the help of some handy technology, those chores become easier to manage.

Laundry in Augury, for example, is more inconvenient in some ways and less in others. For simplicity in plumbing (and lower risk of flooding), few people have a personal washer and dryer. Public laundromats maintained by Poseidon are available throughout the city, with special smart washer/dryer machines that can notify you via your phone when a load is ready or has moved on to drying. After interfacing with the machine for your load of laundry, you can also remotely change settings and decide when to unlock your machine, so you don't have to babysit a load once you put it in. But don't worry—even if the machine loses network connection or has a glitch, you can still get your laundry out with an override access code. They'll even dispense your detergent for you!

Garbage also takes a little extra work, but the results are worth the effort, as we need to minimize waste in our limited space. Garbage is collected by Hephaestus, so to make their job easier, you sort your garbage into compost, metal, glass, paper, and plastic. Compost goes to Demeter, and the rest Hephaestus recycles into new materials. Garbage that can't be recycled or composted is pulverized then mixed into building blocks for new construction.

Work and play are important parts of being a human being, but don't underestimate the importance of rest. As previously mentioned, Augury runs at a pretty laid-back pace. When most of the crucial functions of the city are automated and your livelihood doesn't rely on the pace of the international economy, there's time to stop and smell the roses.

There are some small electric vehicles in Augury—similar to golf carts—but no full-size automobiles and no internal combustion engines. Augury is designed to be a walking city. Stroll around wherever you like, explore, and breathe (and enjoy that the entire

city, by law, is handicap accessible). There are a lot of activities available to help you relax, like community gardening, arts classes, yoga groups, or just sitting on a swing or hammock in an atrium park. Bask in real or artificial sunlight and you might be visited by one of the city's friendly cats, who are cared for by the Animal Control department with automated feeders, fountains, and litter boxes (which provide compost for Demeter). And with some of the fastest internet in the world hardwired in fiber optics throughout the city, you don't have to worry about connecting with the rest of the world—especially since there are frequent conventions, expositions, and other events bringing new and interesting people from around the world to visit your city.

It's also much easier to have peace of mind when you know your health is secure. Mental and physical healthcare services are publicly funded and available to every citizen. Additionally, Augury is population controlled, based on the availability of space, resources, and certain personnel, so you always know there will be enough doctors, nurses, and therapists to take care of you.

There are some unique challenges to adjust to, of course. You can't live underwater without giving up some conveniences. There are no "great outdoors" to escape to unless you go up to the surface; no woods or mountains or lakes. The atriums are big, but at the end of the day, still enclosed—and with limited space, some parts of the city can feel cramped. Some people will have trouble feeling comfortable looking at the endless abyss of the ocean outside their windows. Though Augury has some of the fastest internet speeds in the world, it's still physically isolated and any visitors have to travel overseas and go through a decompression period. You can't easily drive to the next town for a restaurant or a concert if you feel like it, and your access to certain products, locations, and people is more limited. Some who are used to American culture may feel stifled by the regulations and limitations placed upon them in regard to personal freedoms or economic policy.

But there's no place like the city of Augury. With greater regulation comes a stronger social contract, greater safety, accountability and a

sense of deeper social responsibility. The people who live in Augury are dedicated to each other, to working together as a team and a community, and every citizen is a direct member of the legislative body that democratically decides how the city operates. Everyone who comes to live here knows or learns quickly that down here, the priority is health, both as individuals, and as a common people. Having that common goal and mindset makes a world of difference. Plus, how many cities let you look or walk out onto the ocean floor? In the face of the crushing abyss of the ocean, we all work together, not just to survive—but to **thrive**. Few cities on the surface can say the same.

Evil is a weed; it grows from the neglect and apathy in the world. There is still joy to be found in life and in each other. We have to treat our communities like gardens, and intentionally, *lovingly* tend to them every day.

Do not be daunted by the enormity of the world's grief. Do justly, now. Love mercy, now. Walk humbly now.

You are not obligated to finish the work of perfecting the world, but neither are you at liberty to neglect it.

—paraphrase of Rabbi Tarfon, Pirke Avot 2:21

Nihil bonum obstruet
"Let nothing stand in the way of what is right."